S0-BDI-186

PLANETARY EXPLORATION

PLANETARY EXPLORATION

THIRTY YEARS OF UNMANNED SPACE PROBES

Arthur Smith

Patrick Stephens
Wellingborough, Northamptonshire

The author is indebted to the National Aeronautics and Space Administration (particularly the Jet Propulsion Laboratory at Pasadena), Novosti, British Aerospace, General Dynamics and D. R. Woods for illustrations.

First published in February 1988

British Library Cataloguing in Publication Data

Smith, Arthur E.
Planetary exploration: three decades of planetary probes.
1. Space probes—History
I. Title
629.43'53'09 TL795.3

ISBN 0-85059-915-6

Cover illustrations—Front *Earthrise over the Moon, as seen by an* Apollo *spacecraft.* **Back** *(1) Mars as seen by* Viking *on its approach to the red planet* (JPL). *(2) Martian sunset from* Viking I. *(3) Saturn and three of her moons, Tethys, Dione and Rhea* (JPL). *(4) A false colour computer mosaic of views from* Voyager *providing a fascinating portrait of Jupiter's moon, Io* (JPL).

Patrick Stephens Limited is part of the Thorsons Publishing Group, Wellingborough, Northamptonshire, NN8 2RQ, England

Printed in Great Britain by Richard Clay Limited, Chichester, Sussex

10 9 8 7 6 5 4 3 2 1

Contents

Foreword by Geoff Perry MBE

The pleasure of being invited to write a foreword to Arthur Smith's *Planetary Exploration* was somewhat tempered by the realization that, not having written one before, I was not really sure what was required of me.

Most aspects of space research have been a major and growing interest of mine for half my life and I consider myself privileged to have lived through the exciting early years and growth to maturity of the space age. From the time that V-2 rockets fell near my Essex home towards the end of World War 2, I had no doubt that artificial satellites in orbit round the Earth would appear within my lifetime but, even when reading Arthur C. Clarke's *The Exploration of Space* in 1951, I cannot believe that I expected men to orbit the Earth, land on the Moon, and send robots to image all the planets of the Solar System, with one exception, within four decades.

My personal involvement has been entirely with satellites in low Earth-orbit and my first meeting with Arthur Smith was in December 1966 when, as science correspondent for *The Daily Mirror,* he was one of a host of journalists who descended on Kettering Grammar School to cover the story of the announcement of the use and probable location of the third Soviet cosmodrome. Our contacts increased in the years that followed and I came to respect him as a reporter who really understood what 'space' was all about and who shared my own enthusiasms.

When it comes to deep space exploration my first reaction is that I have been a spectator in the crowd, watching from afar through the TV coverage of the *Apollo* missions to the Moon, marvelling at the imagery of the planets, and watching the dramatic pictures as *Giotto* approached the nucleus of Comet Halley. But, on reflection, I have been much more fortunate than that. I have been taken into the 'Moon room' at Jodrell Bank; listened between their transmissions at ITN to the direct feed from *Apollo 15* Mission Control when Scott and Irwin discovered the 'Genesis rock'; been shown the dust impact detector under construction at the University of Kent at a time when it was by no means certain that it would be included in the *Giotto* payload; and visited NASA's Deep Space Tracking Station at Tidbinbilla. My most breathtaking experience must have been at NASA Headquarters after *Viking 1* landed on Mars, waiting for the first vertical strip of the picture to appear.

Currently there is a lull in activity in the exploration of the planets. The *Voyager 2* fly-by of Neptune in 1989 may well be described by Churchill's phrase as 'the end of the beginning' but the *Phobos* mission planned for launch next year will mark the onset of efforts aimed at the eventual landing of human beings on the surface of Mars.

The time is now appropriate for publication of a book which brings together all the various programmes from NASA, the Soviet Union, ESA and Japan to date. In my opinion Arthur Smith has written just such a book and I commend it to the reader as a valuable contribution to the literature of space research.

Geoff Perry MBE
Kettering
September 1987

Introduction

Almost thirty years have passed since the first man-made objects left the gravitational influence of the Earth to begin the exploration of the Moon and the planets. It is true that nothing outside the bounds of science as it was known before the Space Age has turned up in the dark reaches of our Solar System, but in that time some extraordinary discoveries have been made. In particular, interplanetary flights have proved what most specialists long suspected, that we are truly alone as an intelligent species in our own little corner of the Universe, and indeed that Earth has the monopoly of life amongst the nine planets and their numerous satellites that accompany the Sun in its course through the Galaxy.

If planetary exploration has merely confirmed our prejudices about the physical and biological framework of our Universe, some might ask why it is necessary to embark on such a venture, at the cost of very large sums of money. The answer is complex and may be contentious but to the enthusiasts among space explorers there is no doubt whatsoever that it has all been worth it. To begin with, there is the claim that man is a natural explorer who has already spread over the entire land surface of his own globe and it is a logical extension of that urge, to proceed outwards from the Earth. This is a powerful argument which can only be gainsaid by protagonists who wish no more than to stay at home. A second and basically more intellectual argument is that only by a full and exact knowledge of the other planets circling the Sun will we be able to understand the origin and evolution of the whole system and therefore of our own home, the Earth.

It would be obscurantist in the extreme to deny that knowledge of the Earth is an important and even essential matter for mankind at a time when an ever-growing population is putting unbearable strains on its resources. There may never be nickel mines on the asteroids, but geological knowledge gained from planetary exploration could yet help us to understand how best to exploit our terrestrial mineral reserves. Earthly weather studies have also received a boost from the more generalized knowledge gained of weather systems on the other planets. The unmanned probes of the 20th century will surely be followed in the

21st by manned spacecraft following the same well-marked paths. Then and only then will mankind achieve his true destiny—of personally experiencing those parts of the universe which he can today observe only by proxy.

Pioneer 1, the aptly-named US probe which missed the Moon in 1958, began it all. Since that simple piece of hardware, no fewer than 117 more interplanetary spacecraft have, with varying degrees of success, freed themselves from the grip of terrestrial gravity and have sent back pictures and data of bewildering variety which have totally transformed our model of the Solar System. A substantial number of others ignominiously failed to leave Earth orbit or even the Earth's surface in launch failures which were either pictured world-wide in America's case, or, if they were Russian, quietly ignored by the authorities.

There was one thing that they had in common—their dependence on the physical laws of motion and celestial mechanics discovered by Newton in the 17th century. For instance, unlike the spaceships in science fiction, present craft do not fly straight to the target planet. One day, by means of nuclear or electric propulsion, craft may make quick, direct trips to Mars or Jupiter, but at the moment and for the foreseeable future they can only fly in what are known as Hohmann orbits. In its simplest form, this merely requires the craft to be launched in such a way that it goes into a huge orbit round the Sun which takes it to the orbit of the target planet, to arrive at a point occupied by the planet at its time of arrival.

Obviously this means that the launch must take place at a time when the target is at another point in its orbit. The two then rendezvous, brought together inexorably by the gravitational attraction of the Sun and their own velocities. Since we are still dependent on the fairly crude chemical rockets developed in the middle of the 20th century, all the thrust that gives the probe the velocity to break free from the Earth's attraction and enter the Hohmann orbit is imparted in the first few minutes of the flight.

For the rest of the mission, which may last months or years, the craft is coasting through space. If errors have appeared in the trajectory, they can be corrected by small thrusters expelling gas to speed up or slow down the orbital velocity slightly and so make an impact on the accuracy of the flight. Those craft which are designed to land on or orbit their target planet are equipped with larger rocket engines to slow them down in the target's vicinity and so offer themselves for capture by the alien gravitational field. A combination of rocket motors and parachutes of exotic design can then be used to land packages of instruments on the surface.

Although flights to both the inner and outer planets have to begin with fiery rocket launches, the orbital characteristics for the two groups actually differ quite markedly in principle. All of the flights begin from the surface of the Earth. But our own planet is already orbiting the Sun at 105,000 kmh: in order to reach the inner planets, Venus and Mercury, some of this speed has to be cancelled out so that the craft drops into an orbit closer to the Sun than the Earth's orbit. To reach the outer planets the speed of the rocket launch has to be added to the Earth's orbital speed.

Although they are to some extent limited in their ambitions by the specific impulse available in chemical rockets, the mission planners have an ace up their sleeves which enables them to exceed those limits in a spectacular way. The principle of gravitational slingshot manoeuvres has been employed on a number of occasions to enhance a mission by increasing the number of targets which are achievable, most notably in the *Voyager* programme.

It is based on the fact that when a spacecraft flies close to a planet, as long as the trajectory is carefully planned it can be either speeded up or slowed down and the trajectory deflected from its previous course. Neither of the *Voyager* craft, for instance, could have gone on to their triumphant exploration of Saturn without the enormous boost they received from Jupiter's gravitational field (and also the giant planet's own orbital velocity). And without this help *Voyager 2* would have taken thirty years to reach Uranus instead of the eight years and six months that the journey actually took. *Mariner 10* could not have reached Mercury without the slowing influence of Venus.

Paradoxically, the interplanetary probes have never been equipped with the very latest in technology by the time they reached their targets. By the very nature of modern technology which is now changing radically in a time scale of about two years, particularly in computers, a spacecraft designed, like *Voyager 2*, in the early seventies, is totally out of date by the time it reaches Uranus in 1986. Many ordinary individuals now have installed in their homes micro-computers more sophisticated than those which have guided *Voyager 2* through the cosmos.

Naturally enough, these time delays are a direct result of the need to fly in a Hohmann orbit and are proportional to the distance from the Earth to the target. Going to Venus, for instance, is a fairly brief flight. A typical Soviet *Venera* craft takes only just over three months. A flight to Mars, by contrast, because of the much longer Hohmann orbit, is a much more lengthy affair. *Viking 1* took ten months from launch to entering an orbit round Mars.

Months lengthen to years when the planets outside Mars are the targets. To take the most extreme example so far, *Voyager 2* took just under two years to reach Jupiter, almost exactly four years to reach Saturn, and almost 8½ years to pass Uranus, even with the gravitational assist. Understandably, these long flights place great stress on components and instruments and it is one of the technical wonders of the 20th century that by clever design and careful nursing of damaged and worn parts interplanetary probes are kept operating for so long.

To the planets, man has now added a comet to the list of targets attainable by probes. Halley's Comet was a disappointment to most Earth-based observers in 1985 and 1986, but its study by European, Japanese and Russian spacecraft was a triumph of technology and international co-operation. It, too, has a part to play in unravelling the mysteries of the Solar System and perhaps when it comes again after an interlude of 76 years the spacecraft that rendezvous with it will contain the first astronauts to fly close to a comet.

In accordance with international convention in space matters, metric units are used throughout this publication.

	Diameter in km	**Mass × Earth**	**Mean distance from Sun (a)**	**Rotation period**
Mercury	4,878	0.0558	57.9	58.65 hours
Venus	12,100	0.8150	108.2	243.01 hours
Mars	6,796	0.1074	227.9	24.62 hours
Jupiter	143,800	317.893	778.3	10.233 hours
Saturn	120,000	95.147	1,427	10.233 hours
Uranus	52,290	14.54	2,869	17.3 hours
Neptune	49,500	17.23	4,496	15.8 hours

Characteristics of the planets explored by space probes and, in the case of Neptune, to be explored. (a in millions of kilometres.)

Chapter 1

America tries for the Moon

Early attempts to reach the Moon with the somewhat crude space vehicles of the late 1950s and early 1960s had a strong competitive element in them. This was at a time when the so-called 'Space Race' was at its hottest and while the spaceshots were certainly technical they were far from scientific. Getting there was the whole game and any scientific data retrieved on the way was regarded as a bonus and not the whole objective.

It was in this atmosphere that Sir Bernard Lovell, the director of the Jodrell Bank Radio-astronomy Observatory of Manchester University, was drawn into the struggle for technical supremacy between the two superpowers. He recalls in his book *The Story of Jodrell Bank* (Oxford University Press) that just before Easter 1958, he received a secret message from a Colonel in the US Air Force, asking for an interview. The Colonel came to Prof Lovell's office at Jodrell Bank and, mysteriously, asked him to close all the windows and lock the door. Sir Bernard says: 'The real conversation then began in a scarcely audible near-whisper. I was astonished at the message he brought and by his request'.

The Colonel informed him that the USAF had decided to use their Atlas ICBM to launch a rocket to the Moon. They could be ready by August but had no means of tracking the probe as far as the Moon and wanted Jodrell Bank's 250 ft radio-telescope to be used for this purpose. There was a demand for strict secrecy and Sir Bernard felt that he could legitimately track the rocket as part of the collaborative efforts in the International Geophysical Year without informing anyone.

For three months the arrangement remained a complete secret, but then an element of farce entered the scene. As part of the preparations, the Space Technology Laboratory in Los Angeles flew to Lancashire a huge trailer full of tracking equipment and bearing on its side in large letters 'Jodrell Bank USAF Project Able'. An enterprising reporter saw the trailer and the secret was out. At this stage in the Space Race it was decided to give the US Air Force three tries for the Moon and the Army two tries. The objective was to send a payload to within 80,000 km of the Moon's far side. It was in a blaze of publicity that the first attempt was made by the Air Force from Cape Canaveral on 17 August 1958, not, as it happened, with the Atlas as a launch vehicle, but with the smaller 'Intermediate Range Ballistic Missile' or IRBM Thor, with a small Able second stage and an Altair third stage...the whole world witnessed the failure when the rocket blew up eight minutes after the launch.

The cognoscenti at Cape Canaveral began to talk of the 'IBRM'—the 'Into the Banana River Missile', since that Florida waterway was where much of the hardware was ending up at that time. This cruel joke was less than fair to the USAF, since they were attempting to apply a totally new technology to a purpose for which it was not designed. The *Pioneer* payload was a small cylinder with a cone on each end, measuring 45 cm overall. It even had a crude form of infra-red camera with which it was hoped to take far side pictures. And when the Air Force tried again with the Thor-Able less than two months later, on 11 October 1958, they failed to achieve the first Moon hard landing by quite a small margin.

This time, Jodrell Bank picked up the signals from the payload ten minutes after its launch as it rose above the Western horizon. By the evening of the launch day, however, it became obvious that the trajectory was several degrees from nominal and the spacecraft would miss the Moon. It had also failed to reach the required velocity of 10.75 kms and was doomed to fall back towards the Earth before reaching the Moon's orbit.

This was a creditable failure, since the amount by which the rocket failed to reach the desired escape velocity was less than 32 metres per second. The error in the course would have meant that the rocket would have missed the Moon by a trifling 3,200 km, but two days after its launch the payload, by now known as *Pioneer 1*, burned up over the Pacific Ocean on re-entering the Earth's atmosphere.

Pioneer 2, another Thor-Able-Altair payload, failed when it also reached an insufficient velocity and travelled only 1,600 km from the Earth before burning up on re-entry. These spacecraft had all been of the order of 34 kg, which was just about all the Thor-Able could manage, although the Thor was later developed into a highly reliable civilian launcher known as the Delta, capable of launching several thousand kilograms. *Pioneer 3*, which was launched on 6 December 1958, and reached 110,000 km, was a mere 5.9 kg. This was a NASA payload on a Juno II rocket, an adaptation of a US Army Jupiter missile and it was the same launcher which enabled America to reach the orbit of the Moon with a probe that then entered a heliocentric orbit, but not before the Soviet Union had achieved this feat with *Luna 1* in January 1959.

Pioneer 3 was the first of the Army attempts. In addition to the obsolescent Jupiter, the Juno III had a second stage made up of a cluster of eleven Sergeant solid rockets, and third and fourth stages also consisting of Ser-

Right *The Atlas ICBM, converted into a space launcher, was the booster that produced the first successful results in America's lunar and planetary programmes.*

The Atlas-Able 5 payload, which burned up in the atmosphere in September 1960 because of a failure of the Able rocket.

geants—in other words a model of improvization. The first real US success was with *Pioneer 4*, the next Army try, which was launched from Cape Canaveral on 3 March 1959, and two days later passed 60,000 km behind the Moon before entering an orbit round the Sun which intersects the Earth's orbit. It might have been even closer to the Moon in its trajectory if the second stage had not fired for too long.

Pioneer 5, which employed the Atlas-Able combination as a booster, was a much more sophisticated payload which was launched without a specific lunar or planetary destination in mind. At 41 kg it was still far short of the weight of *Luna 1* (362 kg) but it went some way towards calming America's inferiority complex. The 66 cm sphere with the four newly-developed panels of solar cells for a power supply was launched on the Atlas on 11 March 1960. It entered a heliocentric orbit, inside the Earth's orbit, and transmitted telemetry for some time from its four experiments. They were involved with the measurement of cosmic rays over a wide range of energies, ionization and electrical charges in space, magnetism, and micro-meteoroids.

For this mission Jodrell Bank had an active as well as a tracking role. Soon after the launch, when *Pioneer 5* was already 8,000 km from Earth, a signal was sent from the big 250 ft radio dish instructing the payload to separate from its second stage rocket. Contact was finally lost when *Pioneer 5* was 36,000,000 km from Earth. It was the first spacecraft to track the boundary where the Earth's magnetic field gives way to the solar wind, the so-called magnetopause.

This was a success which had to be enough to satisfy the American craving for triumph over the Soviet technocrats, for although there were four more *Pioneers* in the Atlas-Able programme, intended to go into orbit round the Moon, they all failed. One rocket exploded on a Cape Canaveral pad during a static test, two failed during the powered stage of flight, and one exploded in flight.

These early *Pioneer* probes served a useful purpose in that they hastened the learning process of the scientists and engineers who were later to forge such success in programmes like *Ranger, Surveyor, Lunar Orbiter, Mariner, Viking* and *Voyager*. They were the bedrock on which NASA and the US aerospace companies were to build a mighty edifice of lunar and planetary exploration.

Chapter 2
Soviet lunar exploration

When, in the late 1950s, the Soviet Union set out to explore the Moon with unmanned probes it had the advantage over the United States that it had developed what was by the standards of those days a huge rocket. It was an intercontinental ballistic missile in its original form and its design and construction were attributed to the chief designer of the Soviet space programme, Sergei Korolev, although not until after his death in 1966.

However, one man cannot possibly design the whole of a device as complex as an ICBM. It is known, for instance, that the rocket engines which power the standard launch vehicle—the 'A' vehicle as the Americans call it, or the 'R-7' according to its Soviet designation—were designed and developed at the Leningrad Gas Dynamics Laboratory. Undoubtedly many other research centres were involved in the design of other components of the rocket and the payloads, just as American and European projects are shared between aerospace companies and research departments.

Although it was known in the West that Soviet research into large rockets was proceeding, partly with the help of captured German scientists and engineers from the V2 project, when the details became known in the late 1950s they came as a shock. At the time, the Soviet Union had not attained the ability to miniaturize hydrogen bombs, as the USA had and the ICBM necessarily had to be very large in order to be able to carry the unwieldy nuclear weapon over a long distance. It was also inefficient by US standards because the designers had gone for size without attempting the weight-saving exercises adopted, for instance, in the contemporary American Atlas.

The result was a huge missile, 32 metres high, and weighing nearly 500 tonnes when it was loaded with its kerosene fuel and liquid oxygen. US design philosophy was to build large rocket motors and use as few of them as possible on each ICBM. But the Russians adopted a different approach, developing fairly small rocket motors which each had four nozzles of quite a moderate size. Thus, the A vehicle consists of a central core with one of these motors and four strap-on boosters which also each have a motor, so that on take-off the rocket has twenty main nozzles firing, together with a further twelve very small guidance rockets.

The original Soviet space launcher, known to the Americans as the 'A vehicle'. In this picture, Yuri Gagarin is being launched into space in the first Vostok *capsule.*

When the details of this rocket finally became known in the West there was some disbelief among propulsion experts. Both the heavy almost boiler-plate construction and the multiplicity of nozzles were contrary to the best principles of Western rocketry and led to a very small payload for such a large launcher. But the proof of the pudding is in the eating and although the R7 is inefficient it is still in service to this day and is used to launch most Soviet payloads, including the manned Soyuz craft.

Development seems to have been minimal, apart from in the upper stages but it suits the Soviet style of space research, with its emphasis on extremely solid construction, relatively simple instrumentation and repeated launches of similar missions. The US emphasis was on highly sophisticated vehicles, small densely instrumented payloads and relatively smaller numbers of missions. This led them to their distinctive design philosophy of lighter rockets until they built the heavy lifter, the Saturn V, for the manned *Apollo* missions to the Moon.

Sputnik 1 was launched into Earth orbit, using the A vehicle with no upper stage, on 4 October 1957. The impact that it had on the world, particularly on the United States, whose self esteem was badly shaken, was considerable. But a similar shock came on 2 January 1959, less than fifteen months later, when *Luna 1* was launched.

Luna 1, or *Mechta* ('Dream'), was not the first craft to be aimed at the Moon. Attempts had been made by the US with lightly instrumented *Pioneers* in 1958, as described in the previous chapter, but they had all failed in one way or another. *Luna 1* was, however, the first vehicle to reach the vicinity of the Moon. The A vehicle had been adapted by the addition of an upper stage which enabled the payload to reach the Earth's escape velocity of 11.2 kms. The payload weighed 361 kg compared with the *Pioneers'* 6 kg and there was also the upper stage, weighing 1,110 kg, which followed the *Luna* to the vicinity of the Moon. Altogether, while the Americans were trying to catch up with the Soviet Union's undoubted achievements, it was an awesome feat.

Even so, it was not a total triumph, for the guidance was not as good as the propulsion. *Luna 1*, equipped with a minimum of instruments and with the USSR's coat of arms on the outside, was intended to crash onto the lunar surface, so becoming the first man-made object to reach another heavenly body. Instead, it missed the Moon by a hefty 6,000 km. Fortuitously it was given an extra boost by the Moon's gravitational field and was diverted into a 450-day heliocentric orbit in which it remains to this day and will stay perhaps for millions of years.

When the next attempt was made just over eight months later, on 12 September 1959, *Luna 2*, which it is assumed was identical in all respects to *Luna 1*, was successful. It struck the Moon after a very fast journey, 432 km from the centre of the Moon in the northern hemisphere. These two missions had very little scientific content and there is no doubt that in the minds of the Soviet leadership there was an awareness of the propaganda value of successful space shots, especially in non-aligned countries where public opinion might be swayed in favour of the Soviet system by technological triumphs.

A significant pointer to this interest in public relations was that the Soviet Academy of Sciences sent to Jodrell Bank a telex listing the frequencies on which *Luna 2* was broadcasting and also the predicted time of impact on the Moon. This followed a complaint to Soviet contacts by Sir Bernard Lovell that he had been unable to pick up the transmissions from *Luna 1*. True to the British tradition, Sir Bernard went off to play cricket before returning to Jodrell Bank and helping to track *Luna 2* to its impact, less than a minute and a half after the time predicted by the Russians.

But the next Soviet mission, *Luna 3*, was a more serious venture and although the technology was still crude, it was the first spacecraft to make a genuine discovery about the Moon. The Soviet Union called *Luna 3* an 'Automatic Interplanetary Station' (*Avtomatichnaya Mezhdplanetnaya Stant-*

The Luna 2 *spacecraft, which hit the Moon less than two years after the beginning of the Space Age.*

siya). It did not rely on batteries as *Luna 1* and *Luna 2* had, but was powered by solar cells, giving it a longer potential life. It was launched just as its predecessors had been in a direct flight to the Moon but was guided with greater precision so that it looped round behind the Moon's far side which is, of course, never seen from Earth since our natural satellite is locked into an orbit which presents one face to us at all times. It is not the 'dark side' of the Moon of popular speech since it is illuminated by the Sun in the course of each lunar day, just as any other part of the Moon is. The achievement of *Luna 3*, which was not to be exceeded by the United States until the success of the first *Lunar Orbiter* almost seven years later, was to photograph the far side of the Moon on 10 October 1959.

It was not a very good picture but it revealed some features which were confirmed later by the much more advanced technology of the *Lunar Orbiter*. The Soviet Union rather optimistically produced a map of the far side of the Moon based on the photograph, including features such as ranges of mountains which did not actually exist. But one prominent crater, named after Konstantin Tsiolkovsky, the redoubtable Russian pioneer of theoretical space studies, certainly did exist and its spectacular size and appearance produced one of the striking images of the *Lunar Orbiter* missions.

Although *Luna 3* was about the same size as its predecessors it represented

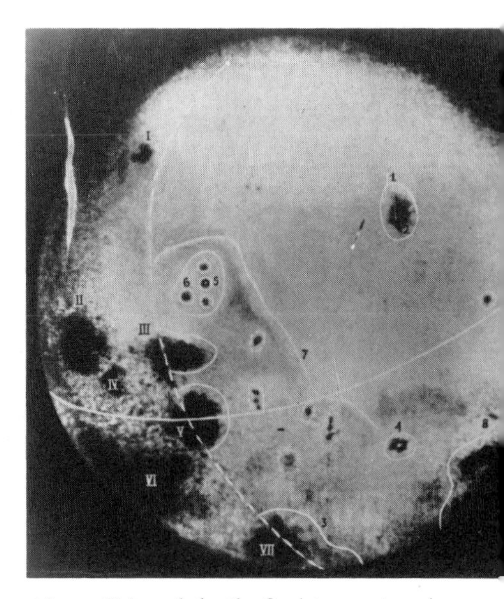

Above *Triumph for the Soviet space team! The first picture ever taken of the far side of the Moon — a major feat by the* Luna 3 *spacecraft. Arabic numerals show real or imagined features on the hidden side, Roman numerals indicate known selenological regions on the side facing Earth.*

Left Luna 3, *which took the first pictures of the far side of the Moon.*

Right Zond 3 *passed within 9,220 km of the Moon in 1965.*

a new step in technology, particularly as it took photography into deep space for the first time. On board was a camera which took the revolutionary photograph of the lunar far side on ordinary photographic film. Then the picture was developed in an automatic 'dark room' and finally scanned by an electro-optical apparatus and beamed back to Earth in digital form. It was an impressively early demonstration of a technique which was also used by America in the *Lunar Orbiter*.

Old-fashioned photography was not to be used much longer in space exploration for as the vidicon tubes which are at the heart of television cameras were perfected and miniaturized it obviously became more sensible to use TV. In such a tube incoming photons are transformed into electrons

which can be fed more or less directly into the transmission device used for sending the pictures back to Earth.

Luna 3 entered what is called a barycentric orbit—that is, one which weaves a rather unpredictable course around both the Moon and the Earth. The fate of such a spacecraft should usually be re-entry into the atmosphere and this is what happened to *Luna 3* after 198 days. The next attempt by the Soviet Union was long-delayed and then it was unsuccessful. It came on 4 January 1963, and it marked a new phase in lunar exploration.

The first three *Luna* craft were launched by a technique which involved a direct trajectory from the Earth to the Moon, the upper stage of the A vehicle firing immediately the core vehicle had burnt all its fuel. This has two disadvantages. One is that there is little time to correct any guidance errors during the all-important period of powered flight. The second is that even the most southerly of the Soviet launch sites at Tyuratam in Kazakhstan is 45.6° north of the Equator and to get the most efficient flight in terms of rocket power it is important to be as near to the Equator as possible at the time of launch.

The new technique, adopted for all planetary and lunar missions by the Soviet Union, involved placing into Earth orbit for at least one revolution the payload and its escape stage, together with a device which came to be called a 'launch platform'. The planners were able to adopt this course because a new upper stage had been brought into service, capable of orbiting 6,600 kg instead of the 4,700 kg of the original upper stage.

Despite the promise of this new technique it did not bring early success. In fact, it failed to deliver a package of instruments to a soft landing on the Moon no fewer than seven times between 1963 and 1965. The first attempt, on 4 January 1963, failed to leave Earth orbit and was given the cover name of *Sputnik 25*. Whenever a payload reached Earth orbit but failed to go any further the Soviet authorities gave it a cover name in this way, most often in the *Cosmos* series of satellites which began to be launched in 1962. When next the technique was attempted, on 2 April 1963, the launch sequence succeeded.

On the first revolution of the Earth, when the spacecraft was over Africa, a firing command was given, probably from one of the tracking ships developed by the Soviet Union, and the *Luna 4* spacecraft was placed on a trajectory towards the Moon. It had a weight of 1,422 kg and was probably a carbon copy of the failed January mission. Unfortunately for the Soviet image, the course was not quite correct and it appears that attempts at mid-course corrections failed. *Luna 4* missed the Moon by 8,500 km. Almost two years elapsed before another attempt was made, and once again a failure was marked up. On 12 March 1965, another *Luna* payload was stranded in Earth orbit and was given the cover name of *Cosmos 60*.

Luna 5 was launched on 12 May 1965, and all seemed well until a few minutes before the projected landing. This time the guidance system worked to perfection but another system, the retro-rocket, let the team down. It failed to work at all and *Luna 5* crashed into the Sea of Clouds at 3 kms.

Luna 6, on 8 June 1965, was another victim of the error-prone guidance system (a symptom of the relatively backward Soviet electronic industry). A mid-course correction failed and *Luna 6* missed the Moon by a dismal 160,000 km. This must have been the low point for the Soviet Moon team—when they corrected one system's faults another would go wrong with monotonous regularity. But they persisted and although there were to be more failures, success was not far away.

It was not to come, however, in 1965. Opportunities for lunar exploration come once every lunar month, or 'lunation', when the Earth-Moon geometry is correct. While they sought to put right the past faults the Moon team had to miss three opportunities but finally launched *Luna 7* on 4 October 1965, which happened to be the eighth anniversary of *Sputnik 1*. This was not, however, a good omen, for although the firing sequence and lunar trajectory guidance were both faultless, everything went wrong again in the last few minutes of the three-day flight. On this occasion, the retro-rocket firing was too early so when the rocket ceased the *Luna* continued to plunge downwards and was wrecked in the Ocean of Storms.

The *Luna* spacecraft must have been coming off an assembly line at this time, for after a gap of only two months yet another attempt was made, *Luna 8*. All was well once again until the very last moment and this time it was a *late* firing of the retro-rocket that caused the probe to crash to its destruction in the Ocean of Storms.

Another two launch opportunities were missed but at last, on 31 January 1966, the Soviet Union scored a triumph which went some way to cancelling America's prestige advantage which was accruing from a series of successful two-man *Gemini* Earth-orbit missions. *Luna 9* was orbited from Tyuratam, the escape stage was fired flawlessly when it was over Africa, and the spacecraft was placed into a perfect trajectory on the way to the Moon. After

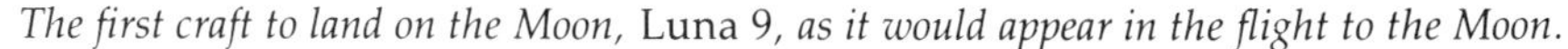

The first craft to land on the Moon, Luna 9, *as it would appear in the flight to the Moon.*

Luna 9's *landing capsule, its 'petals' spread out in a simulated landing.*

a flight of 79 hrs 5 mins *Luna 9* landed in the Ocean of Storms at 18 h 45 m 30 s GMT on 3 February 1966, and so became the first man-made object ever to survive a soft landing on another heavenly body.

The exact landing spot was determined as 7° 8′ north and 64° 22′ west in a part of the Moon that is described as a mare or sea, although this is merely a relic of the belief of ancient astronomers that the dark areas of the Moon are seas. There is no water in the Ocean of Storms or any of the other maria, and *Luna 9* landed in an area of dusty rocks that was quite representative of the lunar lowlands.

The spacecraft as it approached the Moon had a weight of 1,600 kg but most of this weight was made up of the retro-rocket and its liquid fuel. The actual lander that was ejected as the retro-rocket module made a fairly hard landing weighed only 100 kg and consisted of a sphere protected by a shock-absorbing system. It rolled away from the rocket and in a short time four sections of the spherical casing opened outwards like flower petals and stabilized the lander in an upright position.

As the petals opened spring-actuated antennas flipped out and radio transmissions began only 4 minutes 10 seconds after the landing. A television camera was situated in the centre of the sphere and a mirror system above it was able to direct views of the lunar surface down to its lens. It was seven hours before the television transmissions began and some days before the Russians released any of the pictures. But the world was not denied an early look at the rocky moonscape they revealed, for the signals coming back from the Moon were intercepted by Prof Bernard Lovell with his giant radio-telescope at Jodrell Bank near Manchester.

The Jodrell Bank exercise was a model of improvization, for Prof Lovell linked up to the radio-telescope a facsimile machine of the Creed type which was used in those days to transmit pictures between newspaper offices. Although, lacking a calibration scale, Jodrell Bank distorted the pictures, the general effect was seen clearly. It was a rocky scene, with the horizon 1.4 km away. Although the three series of panoramic views taken through the rotating mirror did not reveal anything startling they were a slight comfort to the planners of the forthcoming *Apollo* missions, for they showed no obvious hazards which could make the manned landings impossible. There was, for instance, no sign of the deep dust which some experts had predicted would swallow up the Apollo lunar module.

Luna 9 was powered by batteries and although the tiny 1.5 kg cameramust have had a very small power demand the batteries soon ran down. The his-

Another historic picture — a view of the Moon's surface through the TV camera of Luna 9, *the first spacecraft to land on the lunar surface.*

Luna 10, *which orbited the Moon.*

toric mission ended on 6 February when the deep space control station at Yevpatoriya in the Crimea could no longer raise the faint signals from the Ocean of Storms.

Having this success under their belt, the Soviet team went on to another type of mission—a lunar orbiter. They met with immediate failure when a lunar orbiter was placed into Earth orbit on 1 March 1966, but experienced the same old trouble with the escape stage. Stranded in Earth orbit, from which it decayed only two days later, it was given the cover name *Cosmos 111*. The next attempt was on 31 March when *Luna 10* matched *Luna 9*'s success by embarking on a trip to the Moon which ended with a precise firing of the retro-rocket to place it in lunar orbit.

It appears that *Luna 9* and *Luna 10* were identical in all respects except the instrument package—a contrast with the American solution, which was to build totally different craft, the *Surveyor* and the *Lunar Orbiter*, for the two quite different missions. *Luna 10* beat *Lunar Orbiter 1* to lunar orbit by more than four months, but had a totally different set of mission objectives. The US craft was basically a forerunner of the *Apollo* landings and was designed to take high resolution pictures of landing sites. By contrast, *Luna 10* had no camera and had scientific aims.

It was a 245 kg cylinder and it went into an orbit around the Moon inclined at 71° 54′ to the Equator and varying in distance from the surface by between 1,017 and 350 km. Its scientific tasks were varied, from the detection of meteoroid impacts to the determination of the thermal characteristics of the Moon. A magnetometer was used to detect a lunar magnetic field no more than a thousandth of the strength of the terrestrial field. Studies with a gamma ray spectrometer showed that although there was radiation in belts around the Moon, it was five orders of magnitude less powerful than that in the Van Allen belts around the Earth. This was only to be expected in the absence of the sort of magnetic field which the Earth possesses.

Thermal characteristics of the lunar rocks were measured by means of an infrared instrument and they were determined to resemble basalt more than any other Earth rock. To ram home the political message, the orbiter had been pre-programmed to play back to Earth an electronic version of the *Internationale* on command!

Although *Luna 10* was battery-powered and therefore inevitably had a shorter lifetime than the more sophisticated *Lunar Orbiter*, it nevertheless lasted a creditable 57 days before ceasing transmissions. It could realistically be called the first Soviet lunar craft which returned useful scientific data. Considering the number of attempts and the obvious cost of the programme, *Luna 10* was a measure of the lengths the Soviet leadership were prepared to go to in order to achieve the prestige of technological success.

Four more *Luna* missions were accomplished with the use of the A vehicle with its new escape stage. *Luna 11* (24 August 1966) was placed into a lunar orbit at an inclination of 27 degrees, with altitude varying between 160 and 1,200 km. Transmissions lasted until 1 October and there is a suggestion that television pictures of the lunar surface were beamed back, although none has ever been published.

Twenty-one days after the transmissions ceased, *Luna 12* was launched from Tyuratam, and it went into an equatorial orbit round the Moon on 25 October, its height varying from 100 to 1,740 km. Evidently some development had taken place because *Luna 12* was equipped with both photographic and television cameras and was able to send back an unknown number of pictures of the lunar surface. Even to perform this feat was of considerable value in the learning process for the Moon team, regardless of the quality of the pictures received, for the orientation and stabilization of a craft orbiting a distant body at a range of 400,000 km is a highly sophisticated process. The Soviet *Luna* programme came of age with this mission.

The next payload was *Luna 13*, which was launched from Tyuratam on 21 December 1966, worked perfectly, and landed in the Ocean of Storms slightly further north than *Luna 9*. The concentration of effort on the Ocean of Storms, a large dark area on the left side of the full Moon as you look at it from Earth, suggests that this may have been chosen as a future site for a Soviet manned landing.

Generally speaking, the view transmitted to Earth by *Luna 13*'s TV camera was similar to the earlier pictures from *Luna 9*. A slight difference was that the lander had come to rest with a 16 degree tilt and on one side of the panoramic circle the camera was pointing at the terrain only a metre away. Particles as small as a millimetre could be seen. An experiment, which was new, involved pounding the lunar surface with a pair of explosively-propelled arms, setting up vibrations which enabled sensors on the spacecraft to estimate the firmness and density of the soil.

The results released suggested that the lunar surface at that location was of a material less dense than typical Earth soil, although to a depth of 20 or 30 cm it had similar mechanical properties. The estimated density of 1 gm/cc is that of water and much less than the density of the Moon as a whole, which is about three times as great. Another scientific finding was that the lunar surface reflects about a quarter of the cosmic radiation that falls on it.

Finally, *Luna 14* (7 April 1968) went into a lunar orbit inclined 42° to the Equator and varying in altitude from 160 to 870 km. It was a virtual repeat of the *Luna 10* mission and undoubtedly added valuable data to the body of scientific knowledge of the Moon.

The *Luna* programme using the original two escape stages on the A vehicle had been a remarkable achievement for a nation generally regarded in the West as technologically backward. It had been pursued with an admirable persistence in the face of repeated failures and there was no doubt at all that it had the backing of the Soviet government at the highest level. Although, naturally enough, it had a strong component of military participation because the rocket was a military missile, the planning of the scientific objectives owed a great deal to the Soviet Academy of Sciences and it was therefore basically a civilian programme. If the scientific results were a little sparse the gains in technological experience and reliability as the programme matured were incalculable and were vital to the success of subsequent ventures.

These first two generations of lunar exploration lasted nine years and after a gap of fifteen months the third generation took over. In this new phase the A vehicle was no longer used, although it remained an important part of both manned and unmanned operations in Earth orbit. All subsequent *Luna* spacecraft were launched as the payload of the Proton rocket, or the 'D' vehicle as the Americans call it.

The Proton was so called because its first payload, back in 1965, was a scientific payload of that name in Earth orbit. The big rocket remains a vital part of the Soviet space programme and is used, for example, to orbit the *Salyut* and *Mir* space stations. Its relevance to the lunar exploration programme is that it is much bigger than the A vehicle and can therefore support much more ambitious types of Moon—and planetary—exploration.

The D rocket is comparable in size with the American Saturn 1 which was used during the early test phases of the Apollo programme and also to lift astronauts to the Skylab space station. It stands over fifty metres high, weighs about 700 tonnes as it stands fully fuelled on the pad, and is capable of putting into Earth orbit 20 tonnes or so of payload. This payload, of course, includes an escape stage to propel the scientific instruments out of Earth orbit in the case of lunar or planetary missions.

Like its predecessor the D vehicle consists of a central core with strap-on boosters which fall off early in the powered flight. In the case of the D vehicle there are six strap-ons, not four. Like the A vehicle it is built in the sturdy, boiler-plate style of Soviet tradition and it has so far not been used with high-energy fuels like liquid hydrogen. The success of the US programmes, both manned and unmanned, owes a great deal to the high-energy rockets like the Saturn, the Centaur and later the Space Shuttle, using liquid hydrogen to impart more specific thrust than is available with the conventional fuels such as kerosene and hydrazine.

The Soviet Strategic Rocket Forces, who are responsible for the design and launch of all space rockets, have so far not ventured into the difficult technology of liquid hydrogen. Instead, they have stuck to the tried and tested conventional fuels which are easier to handle. It may be that a Soviet version of the Shuttle, now in the development phase, will be the first craft from the USSR to use liquid hydrogen.

The first use of the D vehicle in the exploration of the Moon led to a series of *Zond* missions but these will be dealt with later since they were quite separate from the *Luna* programme, and probably had more to do with the melodramatic 'race' to land the first men on the Moon. But the first D launch with a *Luna* payload was quite dramatic enough without the added complication of Soviet cosmonauts competing with US astronauts.

It is presumed that *Luna 15* was committed to its translunar flight on the first opportunity after its assembly was completed. These opportunities, or 'launch windows', come only every lunation, when the Earth-Moon geometry is favourable for a low-energy flight. But as it happened, the launch window for *Luna 15* on 13 July was also close to the window for *Apollo 11*, the first attempt to land astronauts on the Moon. The slightly different flight profile for *Apollo 11* meant that its launch date was 16 July and there was some concern about safety among US officials when *Luna 15* went into lunar orbit on 17 July.

Astronaut Frank Borman, who had recently visited the Soviet Union, made a dramatic appeal to Soviet officials for orbital information which would enable NASA to determine whether there was a threat to the *Apollo* from the Russian craft. He was given the parameters of the orbit—inclination 120.5 degrees, altitude 54-201 km—and told that there was no intention to endanger the US crew. There were further orbital adjustments bringing the orbit down to 16-108 km on the day that Neil Armstrong and Ed Aldrin landed in the Sea of Tranquility. There was, of course, no real danger of a collision between the two craft. In theory the two orbiters could have been in the same place at the same time in one of their orbits round the Moon, but the probability was so remote as to be incredible. Even in low orbit round a body as small as the Moon there is a very large amount of space and the two craft were at different inclinations to the equator and at different heights. The *Apollo* mission went on to triumphant success, and the Soviet controllers got on with their own mission.

As it happened, they met with only partial success. *Luna 15* was a soft-landing mission, perhaps intended to pre-empt *Apollo 11* by sending back to Earth a small sample of lunar soil. After it had orbited the Moon 52 times,

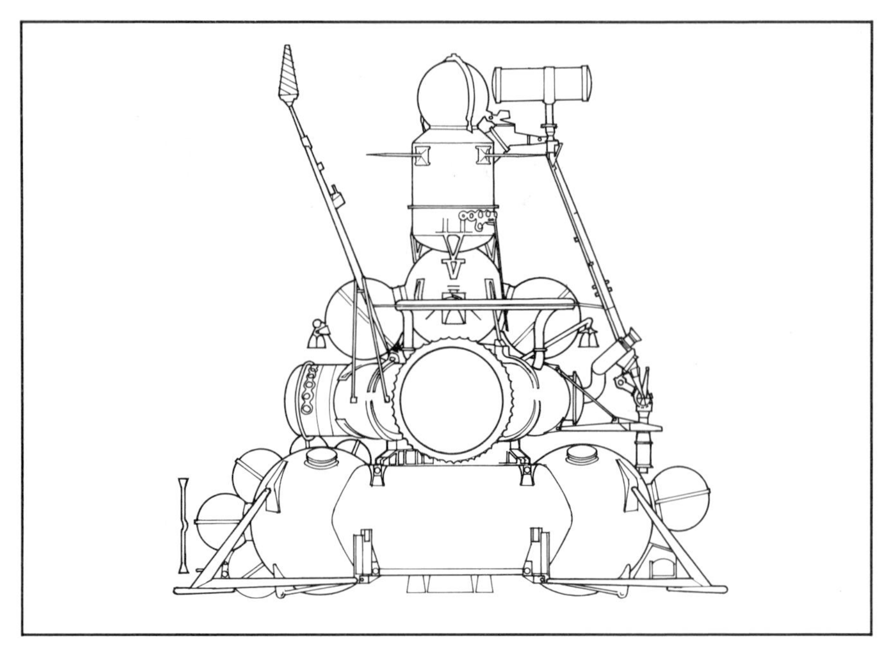

Luna 16 *was the first spacecraft to make a successful unmanned landing on the Moon and return a sample of rocks to Earth. The horizontal cylinder at the top loads the sample into the sphere which returns to Earth.*

Luna 15 was committed to a landing and the retro-rocket was fired. But it was either too late or too early, for Jodrell Bank estimated from tracking data that it struck the Moon at 480 kmh and it was destroyed. Its landing point was both north and east of the position which was occupied at that moment by the lunar module of *Apollo 11*.

Luna 15 was a quantum step forward from the previous generation of explorers. It is estimated that the weight of the craft as it left Earth orbit may have been as high as 6,500 kg. This may have been reduced by 1,000 kg by the manoeuvre to inject it into lunar orbit and to less than 2,000 kg by the landing. Even so, this two-tonne payload was a great contrast with the 100 kg of *Luna 9*, and was large enough to be the base in subsequent years for both sample return and roving vehicle missions. If they had not been overshadowed in the next 3½ years by the far more spectacular *Apollo* missions, these Soviet efforts would have been acclaimed as great technological feats.

The sample return mission which was the next Soviet contribution to the Space Race was undoubtedly planned to upstage the American manned landings in the *Apollo* programme. Unfortunately for the planners the mission was delayed by some of the same old problems and in the event it did not fly until after two of the *Apollo* crews had successfully landed on the

Moon and brought back large samples of lunar rocks. In the meantime, on 23 September and 22 October 1969, two *Luna* attempts were stranded in Earth orbit by the failure of the escape stage and were labelled as *Cosmos 300* and *Cosmos 305*.

Despite these failures and the American successes the decision was taken to go ahead with *Luna 16*, even if its sample would be minute by comparison, and it was duly launched on 12 September 1970. It went into orbit on 16 September and a flawless landing was performed on 20 September. This time Moscow gave more details of how the actual landing took place. By the 20th the low point of the orbit—the 'periselene'—had been reduced to 14 km and when it reached this point the Luna's retro-rocket was fired briefly to lower the orbital speed and so place the spacecraft on a collision course with the Moon. A similar technique was used in the much larger *Apollo* lunar module.

At a height of about 600 m the engine was turned on again by an automatic sequencer combined with radar altimeter data. This burn ceased when the spacecraft was about 20 m up, when two smaller rockets burned to reduce the speed still more. From 2 m the Luna fell freely and safely to the lunar surface in the Sea of Fertility just south of the Equator and on the right hand side of the Moon as viewed from the Earth.

Like *Luna 15*, the lander had a weight of just under 2 tonnes and was a weird-looking device of spheres, cylinders, instrument boxes and struts perched on four stubby legs which absorbed the shock of landing. It stood about 4 m high and, as it turned out, included a descent stage, which stayed on the Moon, and a small ascent stage which eventually returned to Earth. In the 26 hours 25 minutes before this happened a great deal had to be accomplished by the automated lander.

An arm was extended from the lander's descent stage to a region which was unaffected by the exhaust of the landing rockets. At the end of the arm was a drilling bit consisting of a hollow cylinder tipped by cutters. The speed

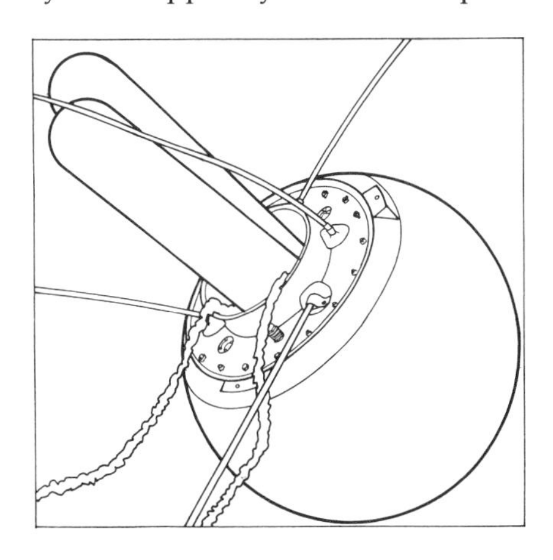

The return module of Luna 16 *after it had descended by parachute to Soviet Central Asia.*

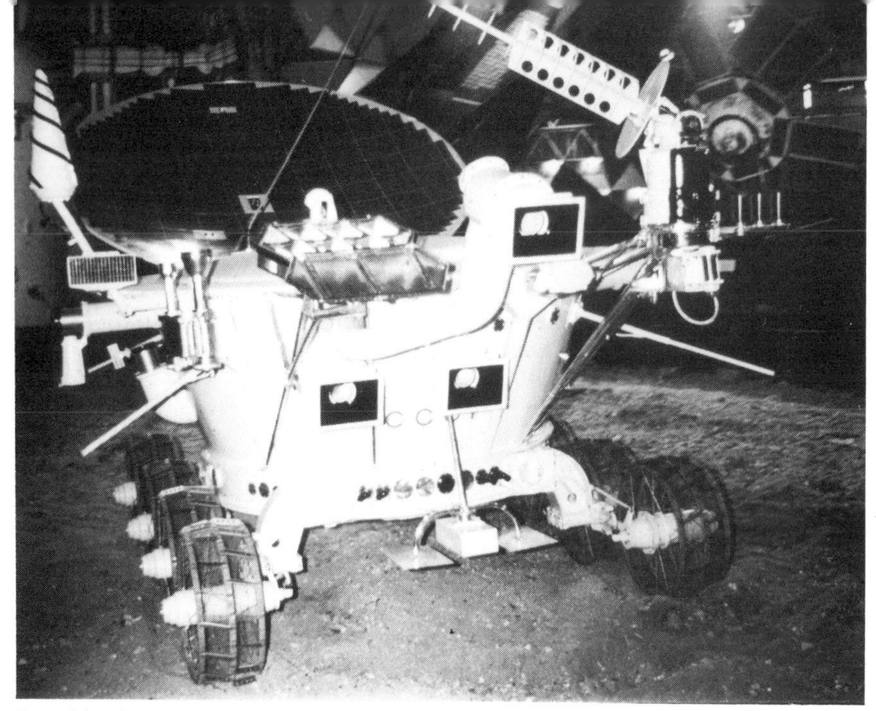

Lunokhod 2, *the improved version of the first roving vehicle to explore the Moon's surface.*

of drilling was monitored by controllers on the ground in the Soviet Union and the drill was allowed to penetrate the lunar soil to a depth of 35 cm. Then it met resistance which may have been bedrock or an isolated stone, but the controllers decided to stop there to avoid damage to the drill.

Then came the trickiest operation so far. Controlled by an automatic programme, the drill cylinder was lifted at the end of its extending arm to the ascent stage above. A container within the sphere which was to be the only part of the spacecraft to survive re-entry into the Earth's atmosphere was filled with the sample and then hermetically sealed. The tiny sample—it weighed only 101 grammes—consisted of grains with sizes typical of sand and some dust.

The ascent stage lifted off on command on 21 September and made the simplest possible return to Earth, that is, a ballistic trajectory with no mid-course corrections and no attempt to use skipping on the atmosphere to reduce speed as the *Apollo* command module did. Such a ballistic re-entry into the Earth's atmosphere would be quite impossible with a manned spacecraft or, indeed, with anything other than the most rugged vehicle. Plunging straight into the atmosphere without any other form of braking, the ascent stage was subjected to 350 times the force of gravity, compared with the maximum of 10G experienced by *Apollo* astronauts on their way back from the Moon. The little sphere was covered with a material which burned away or 'ablated' through the heat of re-entry—a technique which is universally used when it is necessary to recover a payload, except in the case of the Space Shuttle, which uses heat-resistant tiles.

The *Luna 16* sample was studied intensively by the Soviet scientists who revealed that it was grey, although green or brown at some lighting angles, and had a density of 1.8 gm/cc—in other words typical lunar material with which scientists at Houston had already become familiar. The question whether the mission was cost-effective can only be answered in the light of the fact that the Russians were at that time not capable of mounting a manned expedition to the Moon, and may not be even now.

The next best thing was an automated sample return mission and if the early troubles with the escape stage of the Proton rocket had not occurred a sample would have been returned before *Apollo 11*'s success. Small as the sample was, it would have then been hailed as a great achievement and it is perfectly true that the ingenuity and performance of the *Luna 16* concept reflect great credit on the team who designed, built and controlled it.

They soon showed that they still had something up their sleeves because only two lunations later, on 10 November 1970, *Luna 17* was launched from Tyuratam and it turned out to be an even more ingenious mission. The new craft went into orbit round the Moon on 15 November and duly landed in the Sea of Rains in the Northern hemisphere two days later.

As usual, the Russian authorities had not given any advance information about the type of mission, so it came as a complete surprise when it was revealed that in the place of the ascent stage of *Luna 16*, on top of *Luna 17* was an automated roving vehicle called *Lunokhod*—a name which might be translated as 'Moonwalker'. There had been proposals by various American aerospace companies for such a vehicle but NASA had never been enthusiastic about them, clashing as they did with the manned philosophy of the *Apollo* programme.

But in the Soviet Union, it seems, the proposals were taken seriously and some of the ideas put forward may even have been incorporated into the *Lunokhod*. The wire mesh wheels that began to roll in the Sea of Rains, for instance, bore a close resemblance to a design proposed by an American firm. But that does not detract from the achievement of Lunokhod—an idea is one thing, but to put it into practice at a distance of 400,000 km is quite another matter. There were eight such wheels on *Lunokhod*, a strange-looking device which resembled nothing so much as an old-fashioned bath with an opening lid. Its equipment was quite elaborate, including as it did no fewer than four television cameras, solar cells, batteries and scientific apparatus.

Less than two hours after the landing, all its systems checked, the *Lunokhod* started up and descended a steep ramp to the lunar surface. Each of its eight wheels had its own electric motor and the four-man control crew back in the Soviet Union steered by it applying differential power to them. Just in case the controllers attempted to drive it up a slope which was too steep or it tilted too much to one side, automatic sensors on board overruled

Making tracks on the Moon — the impression made by the wheels of Lunokhod 1 *in 1971.*

any instruction which went outside the *Lunokhod*'s capabilities.

The mission went well, with operations being conducted on no fewer than eleven lunar days, each of which is fourteen Earth days long. The power supply came from the lid of the *Lunokhod*, which was covered with solar cells which could be pointed towards the Sun and closed up at night to protect the cells from damage by low temperatures. The scientific apparatus inside the 'bath tub' was protected from damage by a heating system based on a radio-isotope heat source. It included an X-ray spectrometer to analyze soil constituents, cosmic ray detectors and an X-ray telescope to look for X-ray sources in the sky.

Lunokhod 1, a quantum step forward in Soviet lunar exploration, weighed a respectable 756 kg, even though it was made of specially developed lightweight materials. It descended into craters and went as far as 3.5 km from the lander before returning. The Russians were able to pinpoint its location exactly by means of a French-built laser reflector which was fitted on top of the vehicle. With its four TV cameras, *Lunokhod* had vision in all directions and was able to take panoramic as well as close-up and stereo views. Even such a mundane feature as the tracks left by the wheels were studied to investigate the bearing properties of the soil.

Lunokhod 1 was a complete success. It operated from November 1970 to October 1971, and it sent back about 20,000 TV pictures while travelling a total of about 10½ km. It made about 500 mechanical tests of the soil and 25 chemical analyses.

There was more disappointment with *Luna 18*, which was either another *Lunokhod* or a sample return mission. Everything worked perfectly until the touchdown, which the Soviet authorities said was too rough. At any rate, the signals ceased on touchdown and presumably the lander was destroyed. But *Luna 19*, which bore a *Lunokhod* without wheels as a payload, went into lunar orbit on 3 October 1971, five days after it had been launched, and conducted a successful programme of research into the Moon's environment. Magnetic fields, cosmic radiation, solar wind, meteoroids and plasma were studied with the aid of nineteen experiments, and perturbations in the orbit were used to map the concentrations of mass, or mascons, that make the Moon less than a homogeneous spherical body.

Luna 20, launched on 14 February 1972, was another sample return mission. Its target was evidently a repeat of the *Luna 18* aiming point in the Sea of Fertility, where it landed on 21 February. The sample capsule was launched from the Moon on 23 February and returned to Earth, once again in a ballistic trajectory, with a payload of 50 grammes of lunar soil. Analysis showed that it contained no fewer than 70 chemical elements. The sample, which was shared with US and French scientists, had a density of 1.7 to 1.8 gms/cc.

Almost a year passed before another *Luna* mission was launched and this turned out to be another *Lunokhod* flight. *Luna 21* was launched on 8 January 1973, and landed on the Moon in the Sea of Serenity eight days later, only 180 km from the landing site of *Apollo 17* the previous December. *Lunokhod 2* was virtually a repeat of *Lunokhod 1*, but with some improvements and a slightly increased weight at 840 kg. It was able to travel at twice the speed, needed an extra man on the control crew, and yet did not last so long, its life coming to an end after the fourth lunar day in May. During the four months' activity it travelled no less than 37 km on the Moon's surface, including a heroic 16.6 km during the third lunar day in March. More than 80,000 television pictures were transmitted and the rate of mechanical bearing tests on the soil was also increased. More astronomical instruments were carried to study the zodiacal light and the glow in the Moon night sky, and the French laser reflector was used to measure the distance between the Earth and the Moon to an accuracy of within 20-30 cm.

A repeat of *Luna 19*, *Luna 22* was launched from Tyuratam on 29 May 1974, and was placed in a circular orbit of about 200 km inclined 20° to the lunar equator, compared with *Luna 19*'s inclination of about 40°. While it carried out its photographic mission, together with numerous scientific studies, its orbital parameters were changed several times. There was a deftness of touch about the way in which the spacecraft was handled that demonstrated a new sophistication in the mission control centre near Moscow. This was an extremely successful mission, with the controllers sending about 30,000 radio commands to the spacecraft which carried on its lunar studies for fifteen months. The mission came to an end when the manoeuvring fuel ran out in September 1975.

Bad luck dogged the next in the series, *Luna 23*, which was a sample return mission beginning at Tyuratam on 28 October 1974, and landing in the Sea of Crises from lunar orbit on 6 November. It was intended to drill to a depth of 2.5 m but the sampling drill was evidently damaged in unsuitable terrain. This particular variant of the standard *Luna* craft relying on chemical bat-

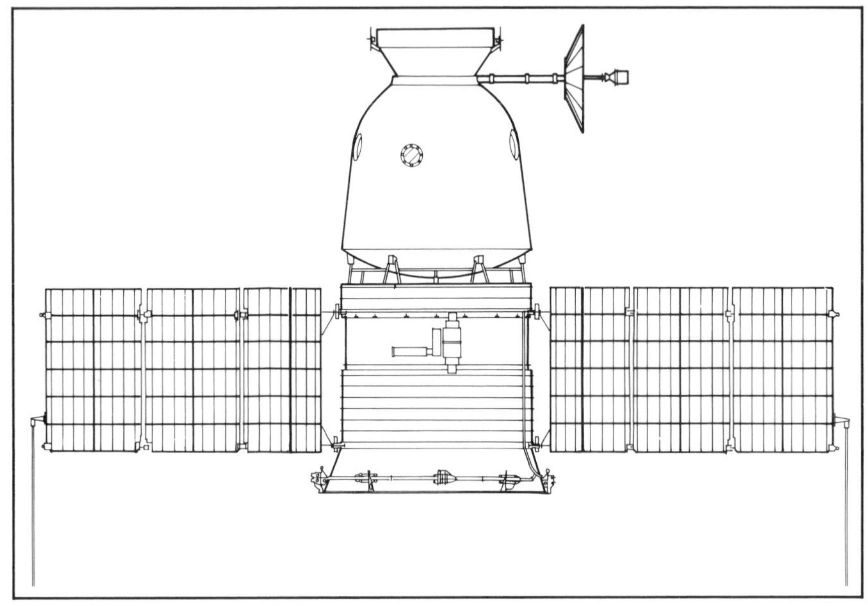

The Zond *spacecraft was a modified version of the Vostok, capable of carrying two men round the Moon and back.*

teries could not last long and the mission was brought to an end on 9 November without a launch of the sample return capsule.

There was then a gap of almost two years before an improved version of the sample return *Luna, No 24*, was launched towards the *Luna 23* target area in the Sea of Crises on 6 August 1976. This was noteworthy for one remarkable fact apart from its complete success—it was the last Moon mission to be mounted by the Soviet Union. At the time of writing, more than ten years after *Luna 24*, the Moon had not been visited again by any Soviet, or for that matter American, craft.

It is well known why the American administration, after the successful completion of six lunar landings in Project *Apollo* in 1973, felt that there was no further point in pursuing lunar exploration. But as far as the author knows it has never been explained by any Soviet authority why the *Luna* programme was cut off in 1976, just at the time when it had become mature and was yielding real scientific data in large quantities. This is one of the most puzzling aspects of the enigmatic Soviet space programme. It may be that it was considered that the basic data needed for a manned programme had been gained and there was no further need for automatic exploration missions. If so, the manned landing has been long delayed.

Luna 24 landed on the Moon on 18 August 1976, and almost immediately deployed a specially-designed drill which was capable of drilling to a depth of over 2 m. An ingenious sytem using a flexible lining for the drill pipe and winching it back into the sample return capsule like a string of sausages successfully recovered the biggest sample yet of 170 grammes of lunar rock. The capsule was returned to Earth, landing in the Soviet Union, where the rocks were analysed and found to contain traces of sixty elements.

This was the end of the *Luna* programme, but running in parallel with it for 2½ years was another programme, which was undoubtedly man-related and has always been so regarded in the West. Between March 1968 and October 1970, five spacecraft called *Zonds* were launched from Tyuratam and four returned to re-enter the Earth's atmosphere after circumnavigating the Moon. It is not a coincidence that NASA decided after the first of these flights to send *Apollo 8* on such a circumnavigation.

The amazing confidence, or bravado, of NASA in so committing the very first manned flight of the Saturn V rocket to such a risky venture owed not a little to the fear that *Zond 4* was a precursor to an attempt by the Soviet Union to send cosmonauts on just such a flight. In fact, the *Zond* missions concluded without such a propaganda coup. But it is a fact that the *Zond* consisted of a lightened version of the Soyuz craft which is still used as a routine ferry to the Soviet space stations. It was lightened by the removal of the forward compartment, known as the orbital module or work compartment, which is jettisoned before re-entry in an Earth orbital mission. So the *Zond* consisted merely of the small re-entry module in which the crew descend to Earth and the service module which includes the Soyuz manoeuvring engine.

The missions were as follows: *Zond 4*—2 March 1968; *Zond 5*—14 September 1968; *Zond 6*—10 November 1968; *Zond 7*—7 August 1969; *Zond 8*—20 October 1970.

Zond 5's *course round the Moon — a planned manned circumlunar flight in rehearsal?*

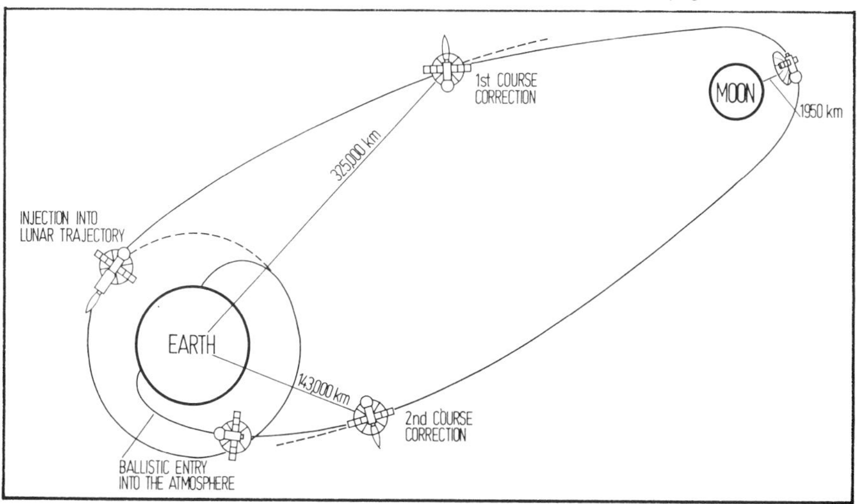

Zond 7 *took this picture of the Moon in 1969 — a remarkable improvement over the* Luna 3 *picture. The numbers mark named craters.*

In the first mission, the Moon was not the target; *Zond 4* was launched towards a notional point on the Moon's orbit. *Zond 5* did go round the Moon with living passengers including turtles, flies, worms and plants and returned after a seven-day flight to splash down in the Indian Ocean, where it was recovered by Soviet ships. Its re-entry was by the direct, or ballistic, path, which would have placed intolerable G-forces on a human crew and incidentally heated the outside of the capsule to a temperature of 13,000°C.

Zond 6 made a more bearable return to Earth by skipping on the top layers of the atmosphere and so reducing the re-entry velocity from the initial 11 km/s to 7.6 km/s before finally entering the atmosphere. This method, which was also used in Project *Apollo,* halved the G-forces which would have been experienced by a human crew. It also enabled the *Zond* to be landed in the Soviet Union in an impressive feat of navigation and control. *Zond 7* repeated this feat, with the addition of colour photography of the Moon and the Earth during the flight. *Zond 8* was a repeat of the water landing in the Indian Ocean with one important difference. *Zond 6* had come up over the South Pole to its watery landing, but *Zond 8* came in by the northern route, over the North Pole, where controllers in the Soviet Union could communicate with it direct for most of its flight.

Nothing came of the *Zond* method of manned lunar exploration, and any Western expert with knowledge of the Soviet space programme expects that when the Soviet Union does resume lunar exploration it will be with a completely new and re-designed system.

Chapter 3
Success for America

The United States planned three programmes to prepare the way for the *Apollo* manned landings on the Moon, *Ranger, Surveyor* and *Lunar Orbiter*. If the early failures of the *Ranger* project had been an indication of the chances of success in all three, they might have been abandoned at an early stage. But NASA persevered and in the end *Ranger*, an unmitigated disaster for 2½ years, proved to be an invaluable precursor to *Apollo*.

Ranger originated late in 1959 as a series of five spacecraft which would take photographs of the lunar surface as they descended to a hard landing. It was also planned that they would release a small instrumented capsule on to the surface. The launcher chosen was the combination of the Atlas ICBM and the Agena upper stage which had been developed for USAF reconnaissance satellite launches. This still did not match the huge Soviet launcher, but it was adequate taking into account the skill at miniaturization that the NASA engineers were developing.

Ranger was a hexagonal spacecraft, weighing up to 364 kg. The electronic systems were inside the hexagonal structure, while solar panels, antennas and experimental booms protruded from it. *Ranger 1* was launched on 23 August 1961, less than two years after the initiation of the project; it was a test flight which was intended to exercise the Agena's ability to re-start in space. The aim was first to place the spacecraft briefly into a parking orbit round the Earth and then to fire the Agena again. This would boost the *Ranger* into a highly elliptical orbit with an apogee of 800,000 km, well past the distanced needed to reach the Moon.

Unfortunately, for unknown reasons, the Agena shut down very shortly after its re-ignition and the *Ranger* stayed stubbornly in its low orbit, from which it decayed into the atmosphere after less than seven days. This was a bitter disappointment, which was compounded when precisely the same sequence of events occurred with *Ranger 2*, which was launched from Florida on 17 November 1961. This time the flight in low Earth orbit lasted only two days.

The next phase of the programme was designed to proceed with the task of landing an instrumented capsule on the Moon and despite the first two disappointing missions NASA decided to press ahead with it. After the

The second Ranger *spacecraft. The balsawood sphere intended for a soft landing on the Moon is at the bottom.*

separation of the capsule from the main spacecraft at a height of about 330 m above the lunar surface, it was expected to impact at a speed of up to 190 kmh. To survive this impact, the capsule was designed as a 30 cm sphere contained inside a 63 cm sphere of balsa wood, weighing 149 kg with its batteries, retro-rocket and other systems. The sphere contained a seismometer designed to record moonquakes or the vibrations caused by the impact of large meteorites.

The last few seconds of the spacecraft's flight would have been hectic, with the capsule separating from the main part of the spacecraft at a height of 24 km only eight seconds before impact, and the retro-rocket reducing the capsule's speed to zero by the time it reached an altitude of 330 m. Unfortunately, the opportunity did not arise to test this sequence of events, for the Atlas-Agena combination placed *Ranger 3* into a trajectory which would miss the Moon by 32,000 km. This was an error too great to be corrected by the mid-course correction engine, and a computer failure made even lunar photography impossible as it passed the Moon.

More disappointment resulted from the *Ranger 4* mission, on 23 April 1962, for the spacecraft apparently began tumbling soon after it had been injected into its lunar trajectory. Solar panels were not deployed either, and the only achievement, if it can be called that, was that *Ranger 4* became the first US spacecraft actually to hit the moon.

Ranger 5, on 18 October 1962, was no more successful. A loss of power from the solar panels and battery problems led to the craft missing the Moon by 725 km. A radical re-think was necessary after five failures and the solution was to fly four more *Ranger* probes purely as photographic missions to

Ranger 6 *successfully reached the Moon but failed to return any pictures.*

support the study of possible *Apollo* landing sites. A major change in design philosophy was to abandon the practice of sterilizing the spacecraft by heat in order to avoid the possibility of contaminating the Moon's environment with Earth organisms. Some of the failures, it was suspected, were caused by the damage resulting from the sterilization. There was later evidence that terrestrial organisms can survive for years on the lunar surface, but the environment is so harsh there that reproduction is unlikely.

Ranger 6 was another failure, however. It carried an impressive package of six TV cameras but no other experiments. Better solar panels and a more powerful mid-course correction engine were fitted. Pictures with a resolution down to 1.8 m were expected to pour back in their thousands during the last few minutes of the flight before the *Ranger* was smashed to pieces on the lunar surface, but although the cameras were switched on eighteen minutes before impact, no pictures were returned. An incident soon after launch, when camera telemetry came on unexpectedly, perhaps through electrical arcing, may have been to blame.

Success came at last with *Ranger 7*, launched from Cape Canaveral on 28 July 1964, and impacting in the Sea of Clouds 68½ hours later. Twenty minutes before its destruction, the craft's six cameras were switched on, and as it accelerated towards the Moon the cameras sent back pictures showing the surface in greater and greater detail. Craters seen in the first pictures expanded until they disappeared out of the frame to be replaced by smaller craters within them. This sequence was repeated several times as resolution improved until, at last, the screens on Earth went blank. The Moon had never been seen in such close-up before and in all 4,316 pictures were received. NASA officials were exuberant—at last the USA had a lunar success to set against the slowly expanding Soviet programme.

It was no fluke, for the last two *Rangers, 8* and *9*, were also completely successful. *Ranger 8*'s target after its launch on 15 February 1965, was the Sea of Tranquility, which turned out in 1969 to be the first *Apollo* landing site. Its 7,137 images of the lunar surface gave a complete picture of the type of terrain—or more correctly 'lunain'—to be found there by the first astronauts to land. *Ranger 9*, on 29 March 1965, was targeted for scientific study of the crater Alphonsus and returned 5,814 pictures to complete the *Ranger* project on a note of high success.

It was the beginning of a period of success for NASA in lunar and planetary exploration. None of the subsequent programmes suffered from the sort of run of failures experienced by the early *Rangers*. There were individual failures, usually associated with the launch vehicles, but the technological experience gained by NASA and its contractors in those early years stood them in good stead as they planned and carried through new missions.

* * *

The next stage in the US exploration of the moon, the *Lunar Orbiter* project, seems ludicrously cheap by later standards. Its five flights, all successful, cost no more than $200 million, including the launch vehicles, which were Atlas-Agenas. In addition to being cheap, it was also a very swift programme, for in May 1964, NASA signed a contract with the Boeing Aerospace Company for the design and construction of five flight models as well as three ground test craft. All were completed and the five flights took place between August 1966 and August 1967—so the whole project took a mere forty months.

Like *Ranger* and *Surveyor*, the project was basically part of the effort to put men on the Moon, and the primary aim of the *Lunar Orbiters* was to photograph at high resolution the potential landing sites for the manned *Apollo* missions, then in the planning stage. For this purpose, the 386 kg craft was equipped with a dual, telephoto and wide-angle, photographic system which used conventional photographic emulsion, developed the pictures and scanned them, sending back the picture elements as electronic pulses. It was the last time anything other than a TV imaging system was used on US planetary spacecraft.

Lunar Orbiter had the appearance of a truncated cone, with a sphere containing the photographic system inside a system of trusses. A velocity control rocket engine surmounted the craft, and it was surrounded by four solar panels. Although it seems simple beside the probes which are now being launched, it was a sophisticated vehicle by the standards of the middle 1960s. After it had fulfilled the fairly limited aims of the manned landing site

Copernicus crater — one of the many fine pictures of the lunar surface taken by Lunar Orbiter 5.

mapping, it went on to photograph more than 36 million square kilometres of the Moon—99.5 percent of the total lunar surface.

Lunar Orbiter 1 took off from Cape Canaveral on 10 August 1966, a year in which a great deal was going right with the US space programmes after the initial setbacks. Lunar trajectory was achieved by the Agena second stage after it had first been placed in a parking orbit round the Earth—one of the rare uses of this technique by the Americans, who preferred the direct ascent method. The craft's velocity control rocket fired as it approached the Moon 92 hours later and placed it into an orbit inclined 12° to the lunar equator with a periselene (nearest point) of 203 km and an aposelene of 1,851 km. The low inclination orbit was chosen because all the potential *Apollo* landing sites were on or near the lunar equator. Periselene was lowered to between 40 and 50 km before the picture sequence began.

Originally there was a list of about thirty landing sites in a band of ten degrees north and south of the equator and between 45'W and 45'E longitude. The *Lunar Orbiter* series, together with *Surveyor*, enabled the mission planners to reduce the sites to eight, each of them an oval measuring about 5 km by 8 km, with an approach path 48 km long for the *Lunar Module*. The first mission was quickly followed by *Lunar Orbiter 2*, on 6 November 1966, and between them the two craft photographed more than 75,000 square kilometres of the landing sites. Resolution in the pictures taken by the 610 mm telephoto lens was down to 1 m, while the 80 mm wide-angle lens pictures showed features down to 8 m in size. Most of the area was photographed again by *Lunar Orbiter 3*, which was launched from the Cape on 5

Many hundreds of pictures of the lunar surface were returned by Lunar Orbiters, *making the journeys of the* Apollo *astronauts a great deal safer.*

February 1967. With the primary objective of the project completed and still two spacecraft to come, new tasks were assigned to *Lunar Orbiters 4* and *5*.

Lunar Orbiter 4, launched on 4 May 1967, had a preliminary lunar orbit of 2,700 × 6,000 km, inclined 85° to the equator and therefore in virtually a lunar polar orbit. From this orbit, and from a lower orbit later it was able to photograph 99.5 per cent of the Moon's near side. *Lunar Orbiter 5*, beginning its mission on 1 August 1967, had a similar orbit, but with a lower periselene, and it was able to complete the photography of the far side of the Moon, which had been started by its four predecessors. After this the orbit was lowered to make it possible to obtain high resolution pictures of five potential *Apollo* landing sites and 36 scientific study sites.

All the orbiters carried instruments to record impacts by micrometeoroids (that is, tiny particles which would be called meteors if seen in the sky as shooting stars, or meteorites if they hit the Earth's surface), and radiation levels. During 800 hours of space operation the orbiters recorded only eighteen hits by meteoroids—a reassuringly low incidence for the *Apollo* planners. Radiation levels were consistently low, except when flares appeared on the Sun's surface.

One bonus for the *Apollo* programme as the first manned flights approached was the fact that the Manned Space Flight tracking network was able to get in some useful practice following the progress of the unmanned orbiters. These operations provided valuable training for tracking station crews and were used to verify computer programs and orbit determinations for Project *Apollo*. The five craft received and executed about 25,000 commands and performed more than 2,000 manoeuvres without a failure.

The fate of the five *Lunar Orbiters* was not to have an honoured place in some museum. To make sure that they did not become dangerous space junk, to interfere with *Apollo* and other projects, it was intended to fire their rocket motors to reduce velocity below orbital level and so cause them to crash onto the surface of the Moon. This was accomplished successfully with four of the craft, but contact was lost with *Lunar Orbiter 4* before the impact command could be sent. However, all was well, for low orbits round the Moon are notoriously unstable because of the gravitational influence of the Earth and the Sun, and *4* eventually crashed naturally five months after its launch.

The film carried on the orbiters allowed only 212 pictures to be taken on each mission. Each of the first two missions returned a very creditable 205.

Lunar Orbiter, *a fairly simple but highly successful concept which demonstrated America's technological strength.*

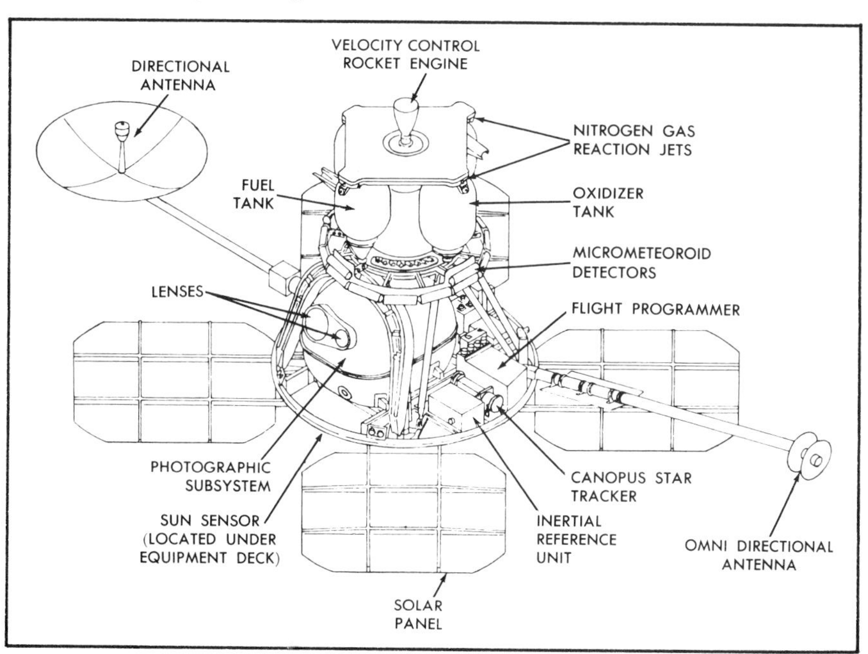

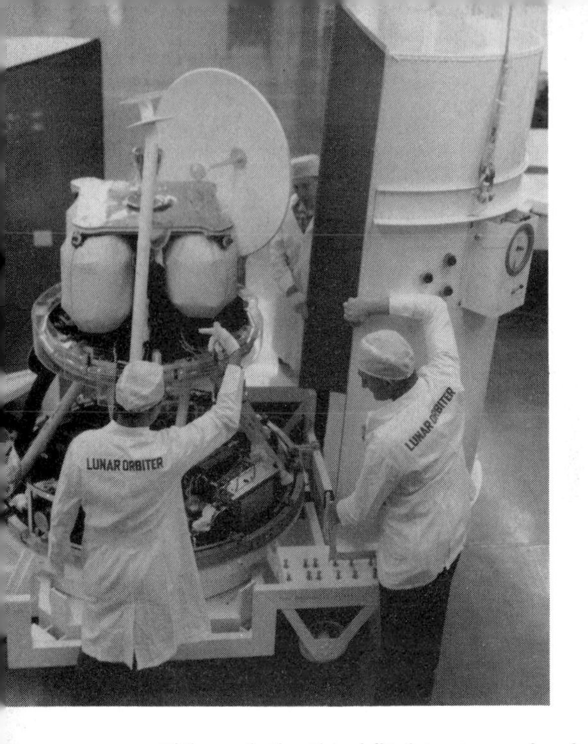

Left *A* Lunar Orbiter *is packed into its travelling container for the journey from the Boeing factory to Cape Canaveral.*

Right *The Atlas-Centaur, ready after many delays, was finally used to send* Surveyor *spacecraft to the Moon.*

Although the third flight returned only 189 and the fourth an even less satisfactory 163, the whole project was vindicated by a perfect performance by *Lunar Orbiter 5*—212 pictures taken and 212 returned to Earth. The pictures were particularly useful in eliminating from the *Apollo* landing site list some apparently smooth areas which turned out to be pock-marked with craters.

Lunar Orbiter 3 was carefully aimed at the crater Flamsteed on the lunar surface with a very special target in mind—the *Surveyor 1* soft lander which had been in this area of the Western hemisphere of the Moon since early in June 1966. The attempt succeeded and *Lunar Orbiter 3* returned to Earth the first picture ever taken of a spacecraft on another heavenly body. This was only one of the several 'firsts' achieved for America by the *Lunar Orbiter* programme.

Lunar Orbiter 1 was the first US spacecraft to go into orbit round the moon. The project also returned the first picture of the Earth to be taken from the vicinity of the moon, it captured the first detailed pictures of the Moon's far side, it gave resolution of the near side ten times better than the best telescopic images, and the first views of the lunar poles. Tracking of the craft also revealed new information about the gravitational field of the Moon, including the existence of 'mascons'—concentrations of mass—under certain areas of the lunar surface. This was important both as fundamental scientific information about the make-up of the Moon, and as operational information for the *Apollo* mission planners.

The *Lunar Orbiter* craft received little public acclaim, lacking as it did the drama of manned participation or spectacular failures. But as a solid, main-

UNITED
STATES
NASA
ATLAS
CENTAUR

stream part of the drive to the Moon it deserves to be seen as a highly successful venture into the field of spacecraft automation.

* * *

The *Surveyor* project, which overlapped with *Lunar Orbiter* in both time and objectives, brought a new level of sophistication into the exploration of the Moon. The *Surveyor* craft were built by the Hughes Aircraft Company (owned by the eccentric multi-millionaire Howard Hughes) and they performed well. Not all of them succeeded but it was encouraging to the team that developed the spacecraft that the first one did. *Surveyor 1* was launched by an Atlas-Centaur rocket, ready at last for escape missions after years of troubles, on 30 May 1966, and landed near the crater Flamsteed on the edge of the Ocean of Storms less than 64 hours later, on 1 June.

Centaur was thus introduced into the lunar planetary programmes of the United States. Unlike all other upper stages used up to that point, Centaur employs the high specific impulse of the combustion of hydrogen and oxygen. This has immense advantages in making large quantities of energy available either to increase payloads or to impart very high velocities to spacecraft. Such propellants, however, are not without their problems: liquid hydrogen has to be maintained at a temperature of −222° Celsius—which places great restraints on the designers of the rocket tanks, as well as on the builders of the valves, pipes and pumps through which the hydrogen has to pass. The Soviet Union has not, as far as is known, developed this difficult technology. NASA incorporated two hydrogen/liquid oxygen stages in the Saturn V Moon rocket and the technology is also employed in the Space Shuttle main engines, but it was the Centaur that started it all and made possible all the earth orbit 'escape' missions after *Surveyor 1.*

Surveyor, America's first soft-landing lunar craft, had three objectives—to soft-land, to provide basic data in support of Project Apollo, and to reveal new scientific knowledge about the Moon. For these objectives it was equipped with a television camera capable of taking pictures of either 200 or 600 lines, and in later missions with means of sampling and analysing lunar soil. At launch the craft weighed 956 kg and on landing, after jettisoning its solid retro-rocket, it had a mass of 270 kg.

The retro-rocket was fired when the *Surveyor* was about 80 km from the Moon, when the lunar gravitational field had already accelerated it to a speed of almost 10,000 kmh. The speed was cut to 400 kmh by the time the spacecraft had descended to 40 km and three small liquid vernier rockets continued the deceleration until the *Surveyor* was dropping at only 4.8 kmh at a height of 4.3 m. The craft dropped freely in the last few metres, so avoiding contamination of the surface by rocket exhaust.

From its landing site, only 14 km from the aiming point, *Surveyor 1* transmitted a total of 10,150 pictures of the lunar landscape during the first lunar

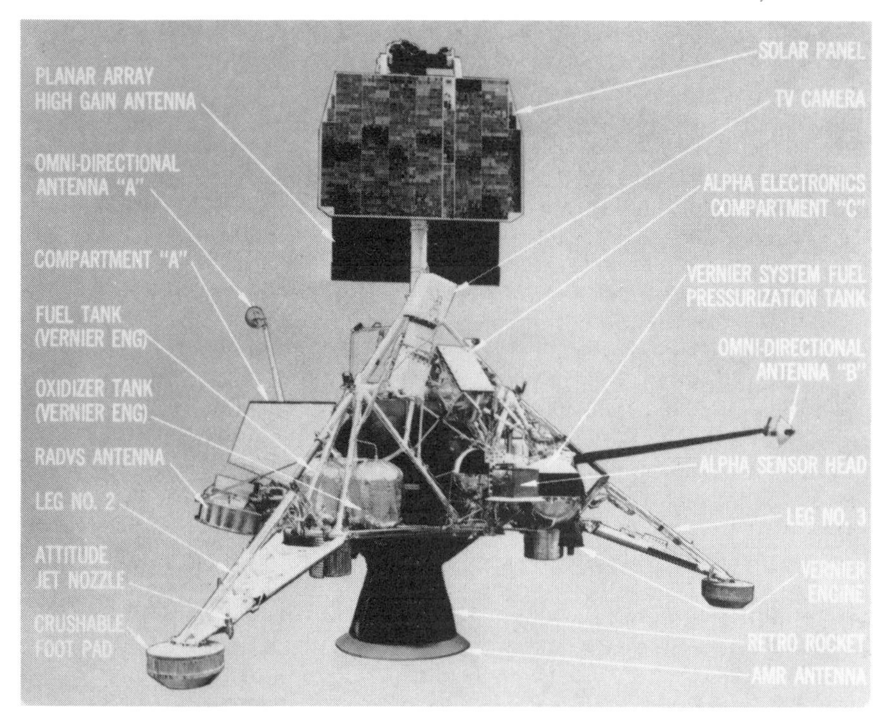

The Surveyor, *later than the Soviet* Luna 9, *but much more complex and in the end more useful as a scientific tool.*

day. A gas jet was turned on to study its erosion effect on the surface, navigational calculations of its position on the lunar surface were made with the aid of observations of Jupiter and the stars Canopus and Sirius, and the solar corona was photographed. Any lander on the Moon has to cope with the extreme conditions of the lunar night, when temperatures drop close to absolute zero for 14 days, 16 hours and 51 minutes. *Surveyor 1* coped admirably with this frigid experience, although it relied on solar panels for its main electric power, and was revived in early July.

It transmitted an additional 1,000 pictures in July and was re-activated several times before finally failing in January 1967, although it sent no more pictures. *Surveyor 1* responded to a total of 158,084 commands from Earth during its lifetime. With this success behind them, the Jet Propulsion Laboratory teams managing the project were disappointed with *Surveyor 2*. Its target was Sinus Medii in the centre of the Moon when it was launched on 20 September 1966, and the Atlas booster and Centaur upper stage both worked perfectly. But during the mid-course correction manoeuvre one of the three vernier engines failed to ignite and the craft began to tumble.

Nothing could be done to correct the tumbling and although the retro-rocket was fired the spacecraft crashed to its destruction at nearly 10,000 kmh near the crater Copernicus.

This was not an omen that *Surveyor* was to be as plagued with failure as *Ranger*, for *Surveyor 3* worked almost perfectly. It was launched from Cape Canaveral on 16 April 1967, and 65 hours later landed in the eastern edge of the Ocean of Storms, just south of the equator and only 4 km from its aiming point. The only deviation from the plan was that the engines did not shut down at 4.3 m because the radar beam controlling the three vernier engines swept over the edge of a crater and they continued to burn right down to the surface. As a result, *Surveyor 3* bounced up to a height of 10 m, touched down and bounced again to 3 m, then landed with the engines shut down and slid laterally for a few centimetres. All the while, fortuitously, the spacecraft stayed upright. Its final resting place was on a 10° slope in a 45 m crater.

Between 19 April and 3 May the craft returned 6,315 high quality pictures of the lunar landscape and of the operation of a scoop-and-claw device which was used to dig trenches, scoop up samples and perform strength-bearing tests by pressing the unit against the soil. It was used to dig four trenches, make seven bearing tests and fourteen penetration tests before lunar night descended on 3 May. This time efforts to revive the spacecraft when the Sun rose again failed. But *Surveyor 3* had not come to the end of its usefulness, for it was used 2½ years later in a unique mission which has still not been matched and perhaps never will be. In 1969 the second manned lunar landing, *Apollo 12*, was targeted on *Surveyor 3* and landed close enough for astronauts Pete Conrad and Alan Bean to walk over to it. They removed parts including the TV camera and brought them back to Earth, where it was discovered that a cable harboured terrestrial micro-organisms which had survived the vacuum, temperature extremes and hard radiation of the lunar surface for 2½ years.

As before, this success was followed by another failure. *Surveyor 4* was launched from Cape Canaveral, with Sinus Medii once again the target, on 14 July 1967. All went well with the mission until the final few seconds of the forty-minute retro-rocket burn. All telemetry abruptly ceased and it is not known whether the automatic landing sequence continued. If it did, *Surveyor 4* will still be standing intact near the centre of the Moon's visible face, or it may have crashed.

JPL's disappointment at this failure was soon allayed by the success of *Surveyor 5*. This was the first in the series to carry a scientific instrument, using the technique of alpha back-scattering, to analyse the composition of the Moon's surface soil. Launched from Cape Canaveral on 8 September 1967, it did not, however, land on the Moon without the intervention of several hours of high drama. Forty hours before touch-down in the Sea of Tranquility it was discovered that there was a leak of the helium which was used to pressurize the three vernier engines that were vital to a perfect

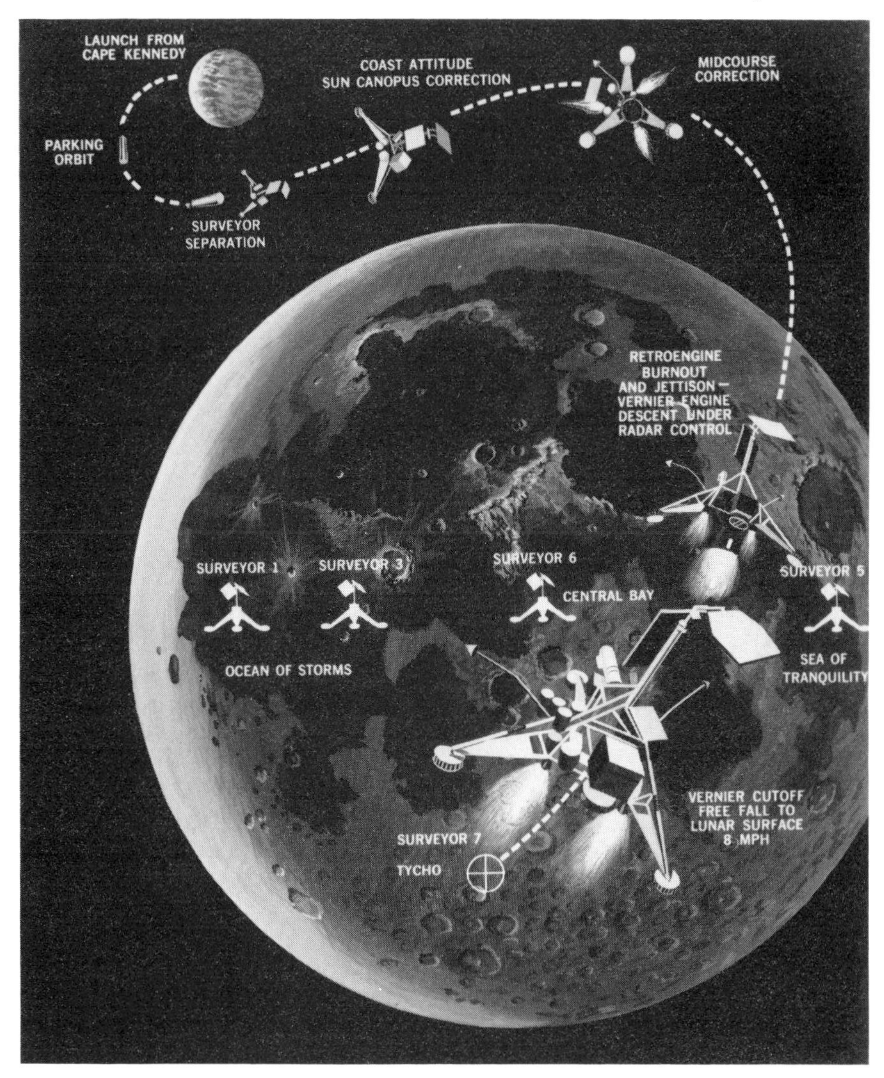

How the Surveyor *craft reached the Moon, and their final resting places.*

landing. It meant that the landing was at risk, for proper ignition of the verniers was essential for completion of the proper sequence.

It was a thousand-to-one against a successful landing in the circumstances, but some improvization was employed to rescue the mission against these impossible odds. It was decided to risk everything by programming the spacecraft so that it would start the retro-rocket burn at an altitude of about 40 km instead of the planned 80 km. Then the vernier engines,

Above *Encounter in the Ocean of Storms. Astronaut Pete Conrad inspects the* Surveyor 3 *lander 2½ years after its touchdown there. The lunar module of* Apollo 12 *is in the background.*

Below *This picture, taken by* Apollo 12 *astronauts, vividly shows how* Surveyor 3 *did a hop, step and a jump when it landed in the Ocean of Storms in April 1967.*

with their reduced pressure, would have to take over only in the last 1,000 m. The hurried calculations made by the JPL engineers proved to be correct and *Surveyor 5* made a safe landing in a small crater less than ten metres in diameter. The slope was 19.9° and one of the three landing legs rested on the rim of a small crater, but the spacecraft was intact.

During the next fourteen days *Surveyor 5* operated perfectly and transmitted a healthy 18,006 television pictures—more than its two predecessors put together. On its second lunar day it sent back another 1,043. Analysis showed that the rock at the landing site was a sort of basalt, similar to terrestrial volcanic rocks.

Surveyor 6 continued the successful sequence taking off from Cape Canaveral on 6 November 1967, and landing in Sinus Medii—the target for the third time—just under three days later. It transmitted 9,741 pictures and conducted soil analysis for sixteen hours before it entered the record books with a unique manoeuvre. On 17 November its vernier engines were fired for 2½ seconds and lifted the spacecraft 3 m above the surface of the Moon and moved it in a westerly direction for two metres. It was the first ever lift-off by a spacecraft from a celestial body. Picture taking was resumed and it was then possible to construct stereo pictures of the original footprints by combining pictures from the first and second sequences. The alpha back-scattering instrument landed upside down at the end of the manoeuvre, thus making it useless for further soil analysis, but it was able to record data on cosmic rays. By the end of the lunar day *Surveyor 6* had transmitted no fewer than 30,027 television pictures.

The last in the series, *Surveyor 7*, was released from the chore of investigating *Apollo* landing sites, since the four which had landed had been so successful. It was targeted to the boulder-strewn region near the Tycho crater 40° south of the lunar equator which is important scientifically because it is overlain with debris from the prominent crater in the form of bright rays. Thus, the rocks in the region should have come from within the Moon, giving the scientists the opportunity to study strata inaccessible from the surface. The craft took off from Cape Canaveral on 6 January 1968, and landed within 2 km of its target 66½ hours later.

It touched down at a speed of only 13 kmh and was soon sending back TV pictures, which ultimately totalled 21,091. It was the only *Surveyor* which carried both the alpha back-scattering instrument and the claw-and-scoop device and they had to be used in conjunction when it was discovered that the former instrument had not been lowered all the way down to the surface. The scoop was moved on to the top of the instrument and pushed it down successfully. The claw was used to test the bearing strength of the soil, which varied from place to place. At one point the claw sank a few centimetres while at another seven times the pressure barely made a dent.

One unique feature of *Surveyor 7* was a laser experiment in which the TV camera took about fifty pictures of laser beams aimed at it from observatories in California and Arizona. The battery of the spacecraft was damaged during the lunar night and operations were limited during the second lunar day, beginning on 12 February. Eventually, the *Surveyor* project came to an end on 20 February 1968, when the craft failed to respond to radio commands. It was the end of an era in which huge strides were made in the understanding of the lunar environment, and it was also the end of the unmanned exploration of the Moon by the United States, for in December of the same year, with the circumlunar flight of *Apollo 8*, the manned expeditions began.

America has chosen for one reason or another not to continue lunar exploration but will be deciding in the next few years whether to resume it. Various options have been suggested, ranging from a manned circumlunar space station to sophisticated unmanned orbiters but there are no firm projects.

Summary of US lunar missions

Name	Date	Result
Ranger 1	23 August 1961	Failed to leave Earth orbit
Ranger 2	17 November 1961	Failed to leave Earth orbit
Ranger 3	26 January 1962	Missed Moon by 32,000 km
Ranger 4	23 April 1962	Hit Moon, no pictures
Ranger 5	18 October 1962	Missed Moon by 725 km
Ranger 6	30 January 1964	Hit Moon, no pictures
Ranger 7	28 July 1964	Success. 4,316 pictures
Ranger 8	17 February 1965	Success. 7,137 pictures
Ranger 9	29 March 1965	Success. 5,814 pictures
Surveyor 1	30 May 1966	Soft-landed. 11,150 pictures
Surveyor 2	20 September 1966	Crashed on Moon
Surveyor 3	16 April 1967	Soft-landed. 6,315 pictures
Surveyor 4	14 July 1967	Landed, but returned no signals
Surveyor 5	8 September 1967	Soft-landed. 19,049 pictures
Surveyor 6	6 November 1967	Soft-landed. 30,027 pictures
Surveyor 7	6 January 1968	Soft-landed. 21,091 pictures
Lunar Orbiter 1	10 August 1966	Success. 205 pictures
Lunar Orbiter 2	6 November 1966	Success. 205 pictures
Lunar Orbiter 3	5 February 1967	Success. 189 pictures
Lunar Orbiter 4	4 May 1967	Success. 163 pictures
Lunar Orbiter 5	1 August 1967	Success. 212 pictures

Chapter 4
The *Venera* programme

After the Moon, the target which has been most attacked by the Soviet space team has been the planet Venus. In addition to the acknowledged missions to the planet, under the programme name *Venera* (the Russian for Venus), there have been a number of unsuccessful attempts to study our nearest neighbour in space. The missions span virtually the entire Space Age to date, for as recently as 1985 balloons and landers were dropped into the Venusian atmosphere by the two *Vega* spacecraft as they passed by on their way to study Halley's Comet.

But it was in much less sophisticated days that the first attempt was made. Not much more than two years after *Sputnik 1* the first use of the launch platform method to send a probe to Venus was attempted. This method, described in the chapter on lunar probes, was to be highly successful in future years, but on this occasion, on 4 February 1961, it did not succeed. The R-7 rocket with a new more powerful upper stage did put the payload of 6,843 kg into a low Earth orbit at a time which was correct for a flight to Venus, but evidently the escape stage which was intended to give the probe itself escape velocity did not work. The whole vehicle stayed in Earth orbit and was called *Tyazheliy (Heavy) Sputnik 4*, although it was announced by the Soviet authorities that this was a test of an Earth orbital platform from which a *Avtomaticheskaya Mezhdplanetnaya Stantsiya* (Automatic Interplanetary Station) could be launched. As in all future Venus missions, a dual launch was planned for this attempt, for *Tyazheliy Sputnik 5* was placed into Earth orbit eight days later, on 12 February. On this occasion the mission came tantalizingly close to success—the escape stage worked and *Venera 1* was sent on its interplanetary trajectory. This probe, the first of a long line of man-made artefacts to be directed at our neighbouring planets in the Solar System, was also the most advanced payload to be demonstrated by either space nation at that time. Its weight was 643.5 kg and its task was almost certainly to strike Venus, thereby becoming the first human artefact to reach another planet. Unfortunately, the weak link this time was the radio. All communications ceased when *Venera 1* was about 7.25 million km from Earth and although it is assumed that it passed about 100,000 km from Venus on 19 May 1961, based on its previous trajectory, nothing is known for certain.

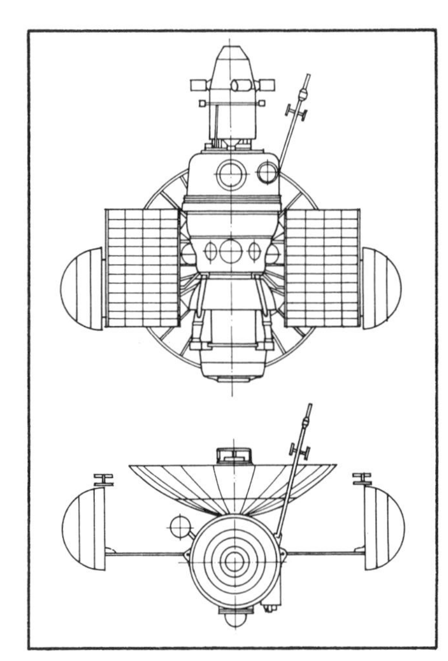

Above left Venera 2, *a partial success for the Soviet Union.*

Above right Venera 3, *which was actually the ninth Soviet attempt to reach Venus, succeeded in striking the planet.*

Launch windows for Venus flights come every 583 days, so it was in the autumn of 1962 that the next attempts were made. At the time the information policy of the Soviet Union on failures had not matured; later there was an accepted formula that lunar or interplanetary probes that did not get beyond Earth orbit should be acknowledged merely as satellites in the *Cosmos* series, but in 1962 there was no such policy. So when three attempts were made and the payloads placed in Earth orbit on 25 August, 1 September and 12 September failed to go any further, Moscow simply ignored them. By this time, of course, the United States had the ability to track Earth satellites and to interpret their purpose and type, thus they were well aware that the three payloads had been orbited and what they were.

There was some confusion at the time because the NASA publication of space payloads, the *Goddard Satellite Situation Report,* omitted the Soviet failures for some time for the sake of tact and the bureaucratic reluctance to reveal secrets, even another nation's! But each launch carries a sequential number by international convention and the Goddard report was published with some numbers missing. This raised some controversy, especially when the Soviet representative at the United Nations accused the US of launching secret military payloads! It was only in 1964 that the mystery was cleared up publicly with a statement by the US about the unacknowledged Soviet failures, but to this day they do not figure in Soviet accounts of space exploration.

The next launch window was in 1964, but before that, on 11 November 1963, the Soviet Union launched what is believed to have been an engineering test flight of the orbital launch platform concept. It was called *Cosmos 21* but it evidently failed too for no probe was launched from it. Undaunted, the Soviet team went ahead with a Venus mission on 27 March 1964, but this too was a failure and was dubbed *Cosmos 27* when its payload stayed stubbornly in Earth orbit. The second part of this campaign was the launch of a further Earth orbital platform on 2 April 1964; this time the escape stage worked well, but curiously, although the payload was obviously on its way to Venus, it was not called *Venera 2* but *Zond 1* (*Probe 1*). This seems to have been an insurance policy against failure—if it didn't reach Venus it could just be passed off as a probe studying interplanetary space. This turned out to be a wise precaution because once again radio communications failed after six weeks and although *Zond 1* reached the orbit of Venus no data returned from it. So far the record was seven attempts and seven failures, no doubt costing hundreds of millions of roubles with no tangible return except possibly engineering experience giving hope for the future.

That hope was to be fulfilled at the next launch window, in 1965, although not without further failures. On 12 November 1965, *Venera 2* was launched from an Earth orbital platform and only 89 days later, on 27 February 1966, it achieved a creditable near miss, passing only 24,000 km from Venus. But there was success in the Soviet programme of planetary exploration finally when *Venera 3*, launched from Earth orbit on 16 November, struck Venus,

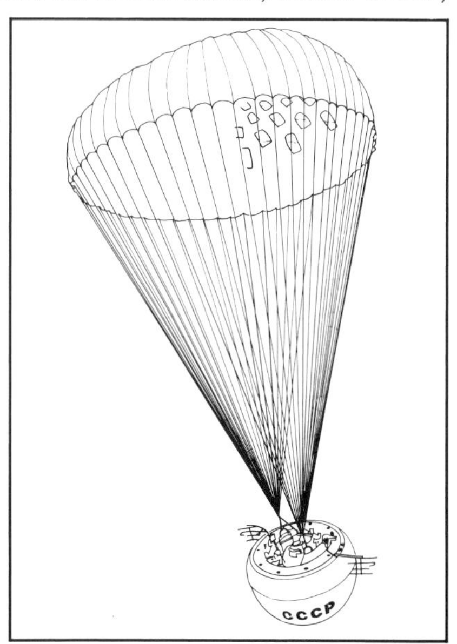

An impression of the landing capsule of Venera 3 *descending through the dense and fiercely hot atmosphere of Venus.*

about 450 km from the centre of the disc as seen from Earth, on 1 March 1966. Propulsion and guidance had at last worked perfectly and although no planetary data had been gathered the Soviet team had demonstrated that once instrumentation and communications matched the other aspects of building a planetary probe they would be able to make a useful contribution to science. The only sour notes were that radio communications had ceased from both the *Venera* craft shortly before they reached the planet, and a third attempt in the same window, on 23 November 1965, failed to leave Earth orbit and was called *Cosmos 96*. *Venera 3* landed a capsule containing a pennant bearing the coat of arms of the Soviet Union on the surface of Venus, ensuring that it was a propaganda, if not a scientific, triumph.

Venera 4, which set off at the next launch window, on 12 June 1967, was a much more sophisticated and successful spacecraft. At this time the Soviet planetary flights were still being accomplished with the R-7 rocket, the original ICBM, plus an escape stage but it seems that some uprating was achieved with the rocket for *Venera 4*, for the payload which went towards Venus totalled 1,106 kg, the biggest so far. This included an instrumented sphere weighing 383 kg which was to travel through the atmosphere and measure its properties.

While the main part of the *Venera* burned up in the atmosphere, the sphere, equipped with a stout heatshield, was braked aerodynamically in the upper layers of the atmosphere on 13 October 1967. When the speed had been reduced by an unspecified amount a drogue parachute was deployed and this pulled from its compartment the main parachute, beneath which the capsule descended for about 1½ hours. At first the Russians thought that they had received data for the whole of the descent until the capsule reached the Venusian surface, but it is more likely that transmissions ceased at an altitude of 25 km.

Venera 4, which deposited the obligatory coat of arms pennant to the Venusian surface, was well instrumented. In addition to two thermometers and a barometer it carried a radar altimeter, an atmospheric density gauge and eleven instruments to analyse gases. The atmospheric gases were sampled at 25 km and 23 km. Data was received for a total of 96 minutes both by the Soviet Union's own deep space network and by Jodrell Bank in England.

These were the first direct measurements from the atmosphere of Venus and were consequently of immense interest to all astronomers who had studied the planet. The first temperature to be recorded was 39°C and the last reading was 277°C. The atmosphere was found to be made up of up to 95 per cent carbon dioxide, with 0.4 to 0.8 per cent oxygen and between 0.1 and 0.7 per cent water vapour. Pressures recorded went up to 22 times the pressure of the Earth's atmosphere at sea level (22 bars), but later spacecraft showed that at the Venusian surface the pressure is up to 100 bars.

The success of *Venera 4*—it survived re-entry temperatures greater than

10,000°—was not matched by its companion in the twin launch of that window. It joined the long line of probes which did not leave Earth orbit when it was launched on 17 June 1967 and was called *Cosmos 167*.

At the next launch opportunity, in 1969, a further pair of craft, *Venera 5* and *6*, were launched successfully, on 5 and 10 January respectively. They were virtually duplicates of *Venera 4* but were more rugged in construction and had smaller parachutes to enable them to sink through the atmosphere more quickly, thus enhancing their chances of survival all the way to the surface. They sent back data for 53 and 51 minutes respectively, suggesting that if they did reach the surface they did not survive there long.

These two spacecraft had a payload mass of 1,130 kg but the next pair, launched in 1970, pushed this mass up to 1,180 kg. On 22 August the second of these two failed to leave Earth orbit and became *Cosmos 359*. But the first, launched on 17 August became *Venera 7* and duly arrived on the edge of the Venusian atmosphere at a speed of 11.6 kms on 15 December. The capsule was successfully deployed and strong signals were received from it as it descended through the atmosphere for 35 minutes. Later computer enhancement was able to decipher a further 23 minutes of telemetry, and showed that the *Venera 7* capsule was the first man-made object to survive on the surface of another planet. This time the true pressure of the atmosphere of the surface was measured as 90 bars plus or minus 15 bars. The surface temperature was recorded as being 748°K, plus or minus 20°.

There were to be only two more attempted Venera missions employing the smaller standard launch vehicle. *Venera 8*, launched on 27 March 1972, was a 1,180 kg payload like its immediate predecessors but they had all been targeted to land on the night side of Venus and *8* was aimed at the day side. It recorded temperatures and pressures consistent with those returned by *Venera 7* but it excelled it by continuing to send back data for fifty minutes after landing. For the first time the capsule included an automatic laboratory capable of making an analysis of the rocks on which it landed. Telemetry showed that the rock at the landing site was similar to granite, with 4 per cent potassium and slight traces of uranium and thorium. The density of the soil was a surprisingly low 1.5 gm per cc.

The other 1972 launch, on 31 March was another failure to leave Earth orbit and was called *Cosmos 482*, and after that there was the first occasion since 1961 that the Soviet Union missed a launch window for Venus missions. They could have gone for Venus in about November 1973, as NASA did with *Mariner 10*, but they were evidently working on a new generation of planetary probes for it was not until 8 June 1975 that *Venera 9* was launched from Tyuratam—and this time the launch vehicle was not the R-7 derived from the ICBM, but the much larger Proton. This big rocket had been used for *Luna* payloads since 1969 so it was introduced to the *Venera* programme at a surprisingly late date. The new craft was

The historic first picture of the surface of Venus, taken by the Venera 9 *lander.*

also innovative in the sense that it consisted of a combined orbiter and landing capsule, and it was also less dependent on Earth commands for its operations, since its equipment included digital computers.

Six days after the launch of *Venera 9*, its twin, *Venera 10*, also left the pad at Tyuratam, so it was a double expedition which set out on a mission which turned out to be the most successful yet in the long Soviet campaign to explore Venus. Two days before *Venera 9* arrived at the planet the capsule separated from the orbiter. On 22 October the 2 m high spherical lander entered the Venusian atmosphere at a speed of 10.7 kms, was braked aerodynamically and by means of a parachute, and landed safely on the surface. The capsule had been built to withstand the pressure of the atmosphere and temperatures up to 2,000°C but even so it survived for only 65 minutes before expiring.

It had been equipped with floodlights but the light was surprisingly good—equal to a cloudy June day in Moscow according to the official announcement—so the lights were not needed. The primary purpose of the lander appeared to be to return television pictures of the surface scene and this it did after a warming up period of fifteen minutes. The vista up to a range of 160 m was seen, with a scattering of rocks of medium size and a large stone on the horizon. The rocks appeared to be young, judging by their sharp edges, and they cast shadows, showing that sunlight was reaching the surface, even through Venus's permanent cloud cover. Venus had been thought to be a geologically dead planet until this mission, but the way in which sharply cleaved rocks were observed demonstrated that there must still be geological activity going on there. The TV cameras on the lander were supplemented by many instruments designed to study the atmosphere and the surface rocks.

Venera 10 arrived three days behind its partner and duly landed successfully. Its landing site exhibited pancake-shaped rocks and possibly lava. The density of the rocks at the landing sites was between 2.7 and 2.9 gm per cc and there were traces of radioactive isotopes of potassium, thorium and uranium.

The two orbiters had periods of just over two days, with high points of 112,000 to 114,000 km. They relayed the signals from their respective landing capsules and were also equipped with instruments to study the planet's outer cloud layers and atmosphere. The orbiters are still in orbit around Venus but ceased transmitting after an operational lifetime of a few months.

Russia missed the next launch window for Venus in early 1977, but embarked on another ambitious dual mission with the departure of *Venera 11* on 9 September 1978, and *Venera 12* five days later. The energy required to reach a planet varies from window to window and on this occasion was greater than in 1975. As a result the strategy was changed so that the orbiter became a fly-by 'bus' and the total weight was reduced from over 4 tonnes to 3,940 kg. By this time the operational side of the *Venera* organization had become mature and the missions proceeded smoothly all the way to the planet. Cosmic rays and gamma radiation were measured *en route*, partly with the assistance of a French-designed experiment, and other instruments monitored plasma, charged particles and the solar wind. On two days in December the plasma instruments were used in conjunction with similar equipment on the US *Pioneer* Venus spacecraft to make synoptic measurements.

Although it had been the second to leave Earth, *Venera 12* arrived first, on 21 December 1978, and its lander entered the atmosphere at 11.2 kms, while its bus flew by at a distance of 32,000 km. The transmissions from

Venera *9 and* 10 *were twin craft, each equipped with an orbiter and a lander which returned soil analyses and the first TV pictures of the surface of Venus* (D.R. Woods).

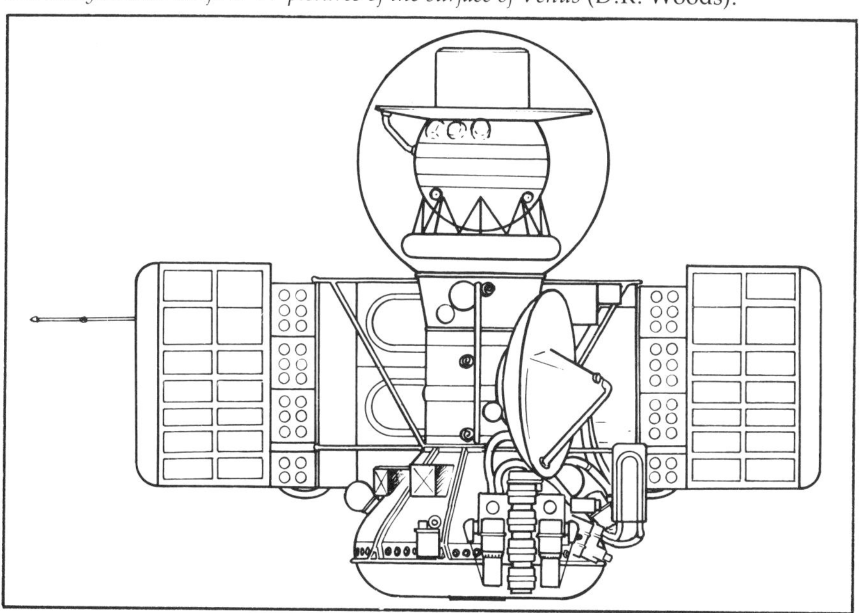

the lander continued for 110 minutes but ceased when the bus, which was relaying them, was occulted by the planet. Four days later a similar sequence of events was followed by *Venera 11*, with 95 minutes of signals coming from the lander. In neither case were pictures received and it is clear that the TV systems on both landers failed to work properly.

Nevertheless, significant amounts of data were received and one interesting fact that emerged is that lightning is a common phenomenon in the atmosphere of Venus, perhaps associated with volcanic eruptions. In one case a thunderstorm region was 150 km across and extended 2 km into the atmosphere. Lightning discharges were far more frequent than in a terrestrial thunderstorm. The landers also found that only 6 per cent of the sunlight falling on the top of the atmosphere reached the surface. The greater sophistication of the instrumentation and communication systems of the new generation of *Venera* craft was demonstrated by the fact that no fewer than 176 mass spectrometer readings were transmitted during the descent of the two landers.

The exploration of Venus continued with the launch of *Venera 13* and *14* on 30 October and 4 November 1981. These were once again 'fly-by bus-plus-lander' missions and each lander took about an hour to descend through the atmosphere four days apart in March 1982, landing about 1,000 km apart in the Phoebus Mountains. This time the difficult task of taking a soil sample for analysis was undertaken successfully in both landers. A sample was passed through an airlock into the pressurized capsule and subjected to irradiation by a radioactive source. This causes the elements to give off characteristic fluorescing radiation which can be detected by a spectrometer. *Venera 13* found that the rocks it sampled consisted of a highly alkaline form of basalt which is rare on Earth, usually found only in oceanic rifts and islands. At the *Venera 14* landing site the rocks were considerably less alkaline and resembled the rocks which make up the Earth's crust.

Colour TV panoramic pictures were transmitted during the brief lives of the landers and showed significant differences between the two sites. *Venera 13* had evidently landed on a portion of the ancient crust of the planet and on a surface that was badly eroded, heavily strewn with crushed fine-grained material. On the other hand, *Venera 14* landed on a smooth surface, clearly stratified and less weathered, and therefore younger. Both sites appeared yellowish in colour, but this was almost certainly because the dense atmosphere on Venus strongly absorbs the blue band of the spectrum.

Wind speed and direction were measurable because some dust was deposited on the landing platform of *Venera 13* during touchdown and was slowly moved by the wind. It was blowing at only about 2 kmh, but that is still a powerful force when the pressure of the atmosphere is 90 bars. A spring-loaded device thrusting a probe into the soil at the *Venera 13* site measured its mechanical properties: the conclusion was that the soil was like a fine-grained, close-packed sand.

At the next launch window, in 1983, a thoroughly redesigned version of the basic *Venera* craft was launched—designed to go into orbit around the

Soviet missions to Venus

Name	Date	Result
Tyazheliy Sputnik 4	2 February 1961	Failed to leave Earth orbit
Venera 1	12 February 1961	Radio failed 4.5 million km from Earth
Unnamed	25 August 1962	Failed to leave Earth orbit
Unnamed	1 September 1962	Failed to leave Earth orbit
Unnamed	12 September 1962	Failed to leave Earth orbit
Cosmos 21	11 November 1963	Failed to leave Earth orbit
Cosmos 27	27 March 1964	Failed to leave Earth orbit
Zond 1	2 April 1964	Radio failed before encounter
Venera 2	11 November 1965	Missed Venus by 23,000 km
Venera 3	16 November 1965	Struck Venus
Cosmos 96	23 November 1965	Failed to leave Earth orbit
Venera 4	12 June 1967	Capsule returned atmosphere data
Cosmos 167	17 June 1967	Failed to leave Earth orbit
Venera 5	5 January 1969	Capsule returned atmosphere data
Venera 6	10 January 1969	Capsule returned atmosphere data
Venera 7	17 August 1970	Capsule landed on Venus
Cosmos 359	22 August 1970	Failed to leave Earth orbit
Venera 8	27 March 1972	Fifty minutes data from surface
Cosmos 482	31 March 1972	Failed to leave Earth orbit
Venera 9	8 June 1975	Orbiter + lander
Venera 10	14 June 1975	Orbiter + lander
Venera 11	9 September 1978	Fly-by bus + lander
Venera 12	14 September 1978	Fly-by bus + lander
Venera 13	30 October 1981	Fly-by bus + lander
Venera 15	2 June 1983	Radar mapped Venusian surface
Venera 16	6 June 1983	Radar mapped Venusian surface
Vega 1	15 December 1984	Balloon + lander
Vega 2	21 December 1984	Balloon + lander

planet and map it with the aid of a radar instrument that would penetrate the cloud cover. Bigger fuel tanks were fitted to contain the fuel necessary to place the whole vehicle into orbit around Venus, the solar panels were increased in size to provide the power needed to run a sideways-looking radar, the high gain antenna's diameter was increased by almost a metre and the transmitters were improved. The result was to increase the data transmission rate to 100,000 bps. *Venera 15* was launched on 2 June and its twin, *Venera 16*, on 7 June. The new craft had been increased in mass to a hefty 5,200 kg and the two craft needed to undergo a slow, 130-day journey to Venus, which they duly reached in October. Their rocket motors fired faultlessly and they entered similar polar orbits ranging from 1,600 to 65,000 km from the planet.

Sideways-looking radar has been used successfully in Earth orbit, notably by the Soviet Union in its programme of ocean surveillance spacecraft. It depends on the fact that a beam of radar waves emitted from the side of the spacecraft and directed downwards at an angle of 10° from the vertical will be reflected by the planetary surface back to an antenna on the spacecraft. The differential return times of different parts of the beam can then be analysed to build up a map of the terrain. In the case of the *Venera 15* and *16* missions the ellipse scanned by the beam was 150 km in width and the resolution was such that features as small as 1.5 km in diameter could be discerned.

Two methods were used to transmit the images of the surface relief thus gained. In the first method, the spacecraft itself correlated the data and transmitted it to Earth stations digitally as instant pictures. The second method consisted of sending the raw data back to Earth for processing at a centre where a higher quality could be achieved. Computer processing of each image took up to eight hours. In addition to the sideways-looking radar the craft carried radar altimeters which were able to measure the height of the surface beneath the ground track with an accuracy of 50 m. The orbits were arranged to give coverage of the sub-polar regions in the northern hemisphere of Venus and the images received, together with results from other research, have enabled the Soviet scientists to publish an atlas of the Venusian topography.

The conclusions reached by the programme of research carried out by *Venera 15* and *16* included the fact that two-thirds of the surface of Venus are covered by hilly terrain, a quarter is flat lowland and a tenth is mountainous. The craft were equipped with infra-red radiometers which pinpointed both cold spots on the Venusian surface, where the temperature was as low as 500°C, and hot spots at 700°C. The first may be areas covered with thick lava sheets, which prevent internal heat from escaping, and the latter may be volcanoes, according to the Soviet scientists. Prof M. Marov, of the USSR Academy of Sciences, believes that the volcanoes are erupting slowly and continuously, not episodically as they more usually do on Earth. Another Academy scientist, A. T.

Bazilevsky, a planetologist, reported the discovery on the radar images of many dome-shaped formations resembling volcanoes. He said: 'It is necessary to find out whether these are ancient volcanoes, which are in great abundance on Earth, or young ones.' He sees Venus as a sort of geological nature reserve, in which can clearly be seen folded mountains, features which look like mid-ocean ridges and ancient ring structures, which are hidden on Earth by kilometres of water or sediments. The northern polar region of Venus, which had previously been hidden from orbiting spacecraft, turns out to be a vast plain, crossed in places by mountain ranges and surrounded by other mountain systems, including the vast semi-circle of Ishtar Terra with 12 km summits.

The exploration of Venus by the two craft yielded 600 miles of the magnetic tape used to record the radar pictures, and it was from these that the Academy produced its map of much of the northern hemisphere of the planet. It consists of 27 sheets, each three square metres in area. The twin mission of the radar probes was a sophisticated and highly professional exercise which showed just how far the *Venera* team have come since they began their endeavours back in 1961. Time will tell whether the *Venera* programme will be halted as the *Luna* programme was, just as it got into a successful phase.

Some of the team were evidently very much involved with the *Vega 1* and *2* mission to pass close by Halley's Comet (described in a later chapter). *Vega* was an improved version of the *Venera*, shielded to protect it from the comet's dust particles, and each Vega craft also deposited a lander and a balloon probe into the Venusian atmosphere as it passed by the planet on the way to the comet rendezvous. This was cleverly done, because the mission planners launched the *Vega* expedition at the very beginning of the Venus launch window. They were able to synchronize the two missions by choosing an unusually long transit time to Venus—virtually six months.

Even so, after the landers and balloons had been released to enter the planet's atmosphere, the rocket motors on the *Vegas* had to be fired in order to manoeuvre them away from Venus, since at that time they were on the same collision course themselves. After passing Venus and acting as relay stations for communications with the landers and balloons, another powered manoeuvre had to be performed to direct the craft to the cometary rendezvous. It was the most difficult and the most complex piece of interplanetary dynamics ever attempted by the Russians and it is a great tribute to their ballistic skill that it was achieved successfully.

Both the landers were able to refine the data from the atmosphere and the surface, and the helium-filled balloons drifted high in the atmosphere transmitting data on constituent gases, temperature and pressure for 48 hours, so expanding by an order of magnitude the time which instruments survived in the corrosive Venusian atmosphere. Almost a quarter of a century of Soviet research into the make-up and dynamics of Venus culminated in this mission.

Chapter 5

America and Venus

America has never shown the intense interest in Venus that has kept the Soviet planners so busy over more than two decades. By the same token, US missions to Venus have not had the same ratio of failure as the long series of *Venera* probes. The select few American missions to Venus have all been successful in returning data except the very first one, *Mariner 1*.

This first representative of a new generation of planetary explorers was originally intended to be a 567 kg payload atop an Atlas-Centaur rocket. Unfortunately, as in so many space projects, the Centaur programme to produce a high-energy upper stage powered by liquid hydrogen was delayed by unforeseen problems and failures in the test launches. *Mariner 1*'s profile had to be altered considerably; it became a 209 kg craft, with only 18 kg of instruments, to be launched by the less powerful combination of an Atlas and an Agena.

It consisted of a basic structure, hexagonal in shape, with a circular high gain antenna for communicating with Earth and a tubular tower bearing an omni-directional aerial. The instruments included a microwave radiometer which would be able to penetrate the dense atmosphere and record emissions from the planetary surface, and an infra-red radiometer sensitive to shorter wavelengths and so able to measure the temperature at the top of the clouds. There was no easy path to Venus, for *Mariner 1* was lost at its launch in July 1962, when the Atlas began to yaw during powered flight from Cape Canaveral. The range safety officer had no alternative but to press the destruct button and launcher and payload fell in pieces into the Atlantic.

There was a back-up, and in due course *Mariner 2* was launched, successfully this time, from Cape Canaveral a few weeks later on 27 August 1962. The launch did not go off without a heart-stopping moment, though, for one of the vernier engines, used on the Atlas for steering, stuck in an extreme position and caused the whole vehicle to rotate round its longitudinal axis once a minute. But the Agena was equal to the task of suppressing this roll after separation and put *Mariner 2* into a temporary parking orbit 185 km above the Earth.

After only thirteen minutes in this orbit the Agena fired again and accelerated the *Mariner* to a speed of 41,544 kmh. Tracking showed that the

Mariner 2, *the relatively simple craft with which America carried out its first exploration of Venus.*

craft on this course would miss Venus by 386,000 km, so a week later, when it was 2.4 million km from Earth the mid-course correction motor was fired and the miss distance was reduced to an estimated 32,000 km. A microphone installed to record hits by micro-meteoroid impacts reacted only twice in seventy days, showing that in interplanetary space there are fewer dust particles than there are in the immediate vicinity of the Earth.

Encounter with the planet took place on 14 December 1962, and, in distinct contrast to some of the later much longer encounters by planetary craft, was all over in 35 minutes (*Voyager 1*'s encounter with the Jovian system lasted 108 days). *Mariner 2*, by then 58 million km from Earth, flew by Venus at a relative speed of 6.7 kms, and less than 40,000 km above the perpetual cloud tops. It was a complete triumph for NASA, contrasting as it did with the dismal Soviet record during the same launch window (three consecutive failures to leave Earth orbit).

The *Mariner 2* microwave radiometer was able to measure the surface temperature at something like 671°K, considerably higher than predicted on theoretical grounds and also higher than the only Soviet reading from *Venera 1*. The mission changed ideas about Venus, for the surface temperature was expected to be about 300°K lower than that recorded. It was also possible to conclude from *Mariner's* measurements that there was little oxygen or water vapour in the atmosphere, a conclusion that was amply proved in later missions. By tracking the spacecraft accurately the JPL

Another view of Mariner 2, *a derivative of the* Ranger *lunar spacecraft.*

scientists were able to produce a more accurate figure for the mass of Venus and also to refine the estimate for the Astronomical Unit, the mean distance between the Earth and the Sun.

The last contact with *Mariner 2* was made when the craft was a record 87 million km from Earth and in a heliocentric orbit in which it remains to this day. It was a triumph for the NASA JPL team and it was only the second time that an American spacecraft had been able to take a long and meaningful look at the interplanetary medium (*Pioneer 5* had been the first).

Five years after the successful flight of *Mariner 2*, NASA resumed its exploration of Venus with *Mariner 5*. This was not a newly designed spacecraft but the spare vehicle from the 1964 Mariner expedition to Mars, which had not been needed because of the success of *Mariner 4*. Engineers at JPL converted it for its new role very quickly. The interval between the NASA decision to go ahead with the mission in December 1965, and the launch from Cape Canaveral on 14 June 1967, must be some sort of record for an interplanetary mission.

Mariner 5, boosted by an Atlas-Agena, left the Earth two days after the

Two views of the Mariner 5 *craft, which passed within 4,000 km of Venus in 1965.*

Soviet Union launched *Venera 4* on a soft-landing mission to Venus, and passed the planet a day after the Russian capsule landed. The Agena's second burn and a mid-course correction burn were precise and when the spacecraft was 79,764,370 km from Earth it passed within 10,151 km of the centre of the planet—this measurement was preferred to a statement of the distance from the surface because at that time the precise 'figure' of Venus was unknown.

By contrast with later spacecraft such as *Viking* and *Voyager* the *Mariner* telemetry was transmitted at a very slow rate—33.33 bits per second for the first forty days and 8.33 bps before and after the Venusian encounter. Even so, more than 3,000 hours of telemetry were received before contact was lost in December 1967. The spacecraft, although relatively simple by later standards, was designed to study the complex interrelations of the atmosphere, ionosphere and plasma environment of Venus. For this purpose, it was equipped with seven experiments, adapted from the *Mariner 4* instruments.

The mission greatly increased our knowledge of Venus. A plasma probe, a magnetometer and energetic particle detectors were designed to study the phenomena which are connected with the way the planet reacts with the interplanetary medium. As in so many planetary missions, the scientists who planned these experiments were interested in the way that Venus responded to the solar wind, the powerful current of plasma that streams away from the Sun in all directions at a speed of about 500 kms. In the case of the Earth, our powerful magnetic field deflects the solar wind so that it never comes closer than 50,000 km, kept at bay by our 'magnetosphere'. But at the time of the launch of *Mariner 5* nothing was known about the plasma environment of Venus.

Mariner 5 passed for two hours within the region around Venus in which the interaction takes place. Its instruments showed that there is an interface between the solar wind and the ionosphere of Venus which is similar to the 'bow shock' ahead of the Earth. But the interaction is not the same, for Venus has a magnetic field which is at least 1,000 times weaker than the Earth's. It could not deflect the solar wind, which is instead deflected by the very dense ionosphere on the daylight side of Venus, which is itself depressed to within 500 km of the planet's surface by the fury of the particles from the Sun. This depression of the ionosphere was demonstrated by one of the two radio occultation experiments.

Meanwhile, an ultra-violet photometer was measuring the properties of the topmost layers of the atmosphere and revealing more hydrogen in the Venusian atmosphere than there is in the Earth's. This is because on Venus the hydrogen is colder than that surrounding the Earth and consequently cannot achieve escape velocity. The instrument confirmed that most of the atmosphere of Venus is composed of carbon dioxide and that in the upper layers, if there was any oxygen, it was too tenuous to measure.

Considering that it was a craft that had been built merely as back-up to *Mariner 4*, *Mariner 5* was an eminently successful mission which produced much new data about our 'twin' planet, as well as taking the art of the operational side of planetary exploration a stage further.

Throughout the entire history of planetary exploration by space probes, spanning almost thirty years, there has been only one mission aimed at the planet Mercury, that elusive ball of rock that is so close to the Sun that few people ever see it. It is quite a bright object, visible to the naked eye, but its orbit near the Sun means that it can only be seen low on the horizon around sunset or sunrise when it has to compete with the twilight.

The reason it has not attracted the same attention as Venus and Mars, the other 'terrestrial,' ie, Earth-like, planets, is simply that it is harder to reach. In order to rendezvous with an inner planet, a spacecraft has to cancel out much of the Earth's orbital speed round the Sun and it has to expend a great deal of energy to do so in the form of rocket power. Thus, more energy is needed to reach Mercury than Venus and the extreme example is the Sun itself. Even the mighty Saturn V rocket which could put more than 100 tonnes into low Earth orbit, would have had a zero payload in a mission to the Sun.

Mercury, then, is a difficult planet to reach, but there is a way round this dilemma, and it was pointed out by a graduate student in Los Angeles, Michael Minovich, as long ago as 1963. He realized that although an Atlas could not reach Mercury with a useful payload on a direct flight, it could do so with the aid of the gravitational field of Venus. This 'gravitational slingshot' technique has been used to good effect in several US missions since then. When we come to the story of the *Voyager* missions it will be seen that it is a highly valuable technique which renders possible some things that would otherwise be impossible and enhances some which would be otherwise less valuable. In the case of *Mariner 10*, which was planned on the basis of Minovich's insight, there were several advantages. The first was that as the gravitational field of Venus was being brought in to assist the flight to Mercury, the Mariner had to fly close to the mysterious planet and could therefore study two planets in one flight. The other main gain, which was discovered during the planning stage, was that after passing Mercury on its elliptical orbit round the Sun, the spacecraft would come back again at least twice in a form of resonance with Mercury's orbit to make possible further studies of the planet.

Mariner 10 was duly launched from Cape Canaveral on 2 November 1973, and after a flight of 95 days encountered Venus on 5 February 1974. Since its main objective was to study Mercury, the trajectory and the aiming point at Venus were dictated by the need to use the Venusian gravitational field as a 'third stage' to reduce the orbital speed. Even so, there were many objectives in the encounter with Venus despite the fact that the instruments carried on the 502 kg spacecraft were optimized for Mercury studies. The objectives included further study of the planet's interaction with the solar wind, a key factor in the environment of any planet.

Radio science was to be used to study the ionosphere and the neutral atmosphere by means of occultations: it was desired to penetrate the scattering layer of the atmosphere observed by *Mariner 5* at an altitude between 35 and 50 km. Extreme ultra-violet emissions were to be studied from hydrogen, helium, oxygen, carbon, neon and argon to provide new information on the composition and structure of the upper atmosphere. There were also to be studies of cloud-top brightness and limb (planet edge) darkening, and the close fly-by of the planet was expected to provide refinements of the ephemeris, mass and figure of Venus.

Imaging of the surface of Mercury required 1.5 m focal length optics for the two television cameras, but these gave a field of view of only one half by one third of a degree. This was too small to examine the visible surface of Venus for clouds, so wide angle auxiliary optics were added to the telescopes to give a 14 × 11° field. The imaging objectives on encountering Venus were to search along the terminator for visible clouds during the few minutes near periapsis when the viewing geometry was ideal, and to study the ultra-violet markings to give important information on upper atmosphere dynamics. For the latter purpose the camera systems were instrumented to match the near-UV wavelength at which the markings had previously been observed. To improve the results near-UV filters with a bandpass between 300 and 400 nanometers were incorporated, and special UV-transmitting optical coatings were added to the corrector plates of the telescopes. The extreme-UV airglow spectrometer was designed to be fixed to the body of the spacecraft for viewing of Mercury only, but it was decided to fix it to the scan platform so that it could be directed at Venus as well.

In many ways this was the most elaborate US planetary explorer built to date. All the experience of the Jet Propulsion Laboratory was put into designing a craft which was stabilized by cold nitrogen jets, guided by sensors which were directed at the Sun and the star Canopus, and powered by two solar panels which could be rotated to reduce solar heating as the orbit moved sunwards. A nickel-cadmium battery supplied power when the panels were off the Sun. With a 1.37 m parabolic high-grain antenna the craft was able to send back data from the region of Venus at a rate of 117,600 bits per second for TV and 2,450 bits per second for science. A small computer and sequencer were updated periodically to control successive portions of the mission, since the delay in transmission by virtue of the distance of the spacecraft from the Earth made real time control impossible—the constant dilemma of scientists exploring planets by proxy.

The scientific instruments were turned on a week before *Mariner 10* reached Venus, when the encounter operations began. At encounter Venus was 44.1 million km from Earth and the spacecraft was travelling at 11.03 kms relative to Venus, which duly did its job by slowing the craft down by 4.41 kms, so making the encounter with Mercury possible.

There was at least one failure during the interplanetary flight. Shortly after the launch it was found that the heaters which were supposed to warm the TV optics would not switch on. The cameras were used for calibration tests by photographing the Moon but then the TV instrument power had to be left on to ensure that the optics did not drop below a critical temperature. This meant that the post-encounter televising of Venus had to be cut short because of concern over the instrument lifetime. But, even so, photography began 38 minutes before encounter and continued for several days as the spacecraft swung away from Venus. The infra-red radiometer made possible observation of the atmosphere at a wavelength inaccessible from Earth. The plasma science experiment showed the interaction between the Venusian atmosphere and the solar wind resembled that thought to occur with a comet in some ways; Venus probably has a tail like a comet, equal in length to hundreds of its radii.

Just putting a trackable object in the near vicinity of a planet provides valuable information, since its deviation from its path brought about by the planet's gravitational field can be detected by Doppler tracking. *Mariner 10* showed in this way that the mass of Venus is such that it would require 408,523.9 of the planet to equal the mass of the Sun. The 'radio science', by tracking the course of the radio waves from *Mariner* through the atmosphere, also showed that there are four temperature inversions there, at 56, 58, 61 and 63 km from the surface.

Another important result was that *Mariner 10* helped to resolve the strange relationship that exists between the rate at which Venus spins and the gravitational attraction of the Earth. Before the Space Age the 'day' of Venus—the period during which it completes one rotation—was unknown because no surface feature which could be tracked was visible through the clouds. It was even thought for many years that just as the Moon presents the same face to the Earth at all times because it is locked into a permanent terrestrial embrace, so Venus always turned the same face to the Sun.

This would mean that the Venusian day was the same as the year, which is 224.7 days. But early missions had shown that Venus actually rotates in about 243 days and in a retrograde motion—that is to say clockwise when viewed from above the north pole. The Earth and all the other planets except Uranus rotate in a counter-clockwise direction. The interesting thing about the rotation of Venus is that if it is exactly 243.16 days, when it is at inferior conjunction (between the Earth and Sun) it does always present the same face to us. *Mariner 10* was able to establish that the rotation period is within one part in 2,500 of the required value. The exact value is now put at 243.01 days, so the relationship between the Earth and Venus is now regarded as a pseudo-synchronization rather than a genuine one.

After its swift visit to Venus, *Mariner 10* continued on its way. After a mid-course correction manoeuvre on 16 March to reduce its miss distance from 10,000 km, it made its closest approach to Mercury, an historic first, on 29 March 1974, at the end of a 146-day journey. Its new course was precise and it passed within 700 km of the planet's dark side. It had not been idle during its cruise between the planets; it continued to collect and relay back to Earth data on magnetic fields, plasmas and charged particles. At encounter, the probe was 148.6 million km from Earth and the spacecraft's speed relative to the planet was 11.3 kms. The encounter phase actually began on 23 March when the first TV pictures were returned at a range of 5.3 million km. From then until sixteen hours before the closest approach, there was an hour of picture taking every day. At this stage commands to the cameras and science instruments were issued from Earth. After T minus 16 hours and for the next 32 hours all the science commands were issued by the spacecraft's own computer, which had been programmed to do so before the far encounter sequence began on 23 March. This was an interesting variation on spacecraft control, to be compared with the later *Voyager* craft, which were so far away from the Earth that direct control was impossible and all sequences of activity were beamed up to the spacecraft in advance.

The pictures at full resolution flowed back for 7 hours 48 minutes through the big dishes of the Deep Space Network at Goldstone in the Californian desert and Tidbinbilla in Australia. Mercury turned out to be a heavily cratered planet, with many resemblances to the Moon. There were extensive areas of terrain morphologically similar to the lunar maria and the surface was generally primordial. It seems that the surfaces of the Moon and Mercury were formed by similar sequences of events, with the consequence that the chemical and mineralogical nature of the outer layers of the two is similar. Mercury is denser than the Moon and this might have suggested a surface history that was unique, but the resemblances to the Moon's surface were extraordinary.

Nevertheless, Mercury is a differentiated planet—that is, the heavier materials have sunk into an iron-rich core, just as they have on Earth. Like all the planets and satellites in the Solar System it was subjected to an intense bombardment by small solid bodies, the 'planetismals', at the time of the formation of the system 4.6 billion years ago. Unlike other planets at the end of this period, it seems to have been subjected to a unique planetary compression which has left scars in the form of large-scale scarps and ridges.

The lithosphere, the outer layer of the planet, was under compression, apparently because the molten core cooled down, and this led to a shrinkage of the planet estimated at 10 km, not a lot with the planetary radius being 2,439 km, but enough to cause 'wrinkles' which can be up to 500 km long and 3 km high. This all happened about 4.6 billion years ago, when the planet formed. There seems to have been another episode of bombardment about 3.9 billion years ago, and then, about 3.6 billion years ago there was another cataclysmic event which left Mercury with its most prominent feature.

This was a collision with an object that must have had a mass of many billions of tonnes and dimensions of many kilometres. It left as a legacy of its destruction a mighty crater 1,300 km in diameter which then became flooded with lava as a result of the volcanic eruptions its own appearance caused. It was photographed by *Mariner 10* half in light and half in darkness so only part of it has been seen. Nevertheless, it was clear enough that this was a major component of the Mercurian surface and it was named the Caloris

Basin. The name was chosen because the basin is on one of the points on the surface which is particularly hot becaause of the strange nature of the planet's relationship with the Sun.

At the Antipodes of the Caloris Basin there is an area of jumbled and broken terrain and shifted craters. When the asteroid or planetismal that caused Caloris hit Mercury the seismic waves must have swept through the planet with devastating force, breaking up the antipodean surface. Similar events may have occurred on Earth in the distant past, but erosion and sedimentation have wiped out almost all traces.

After its encounter with Mercury, *Mariner 10* entered a final heliocentric orbit in which it revolved round the Sun in exactly twice the orbital period of the planet, which is 87.97 days. This meant that after every two Mercurian orbits the spacecraft encountered Mercury again and for two of these encounters Mariner was in working order. At the two encounters, on 21 September 1974 and 16 March 1975, the cameras were able to photograph once again the areas originally imaged at the first encounter. This was because the planet has a 'day' or rotation period of 58.65 Earth days, exactly two-thirds of its year and therefore rotates round its axis exactly three times while going round the Sun twice. This meant that there was a fairly limited coverage of the planetary surface—only 35 per cent was mapped but by chance this included the most interesting feature, the Caloris Basin, and also gave the mission team a good grasp of the essential appearance of the surface.

Because of the resonance of the day and the year, to an observer on the surface of the planet the Sun would appear to linger for a period of eight Earth days over the subsolar point at 'midday' while describing a strange loop in the sky. At this point the temperature reaches an incredibly high temperature of over 700°K (430°Celsius). On the other side of the globe, at 'midnight' the temperature plunges to not much more than 100°K (-170°C), a difference of 870°K. Like so many other of the bodies visited by the twentieth century's incredibly clever robots, Mercury turned out to be an even more weird world than anybody had suspected from earthbound telescopic observations.

With its close proximity to the Sun, Mercury was not expected to have an atmosphere. Molecules of gas can only be bound to a planet in an atmosphere if the escape velocity of the planet is higher than the speed of the molecules in their random motion. The lightest gas, hydrogen, for example, has escaped from the Earth's atmosphere because at a sufficiently high temperature its molecules exceed the Earth's escape velocity (11.2 km/s). But Mercury's escape velocity is not much more than 4 km/s and with the high subsolar temperatures it was expected that all the gases in its primordial atmosphere would have disappeared billions of years ago. Yet the Mariner's instruments did detect a tenuous atmosphere of helium, with a pressure of only an infinitesimal fraction of the Earth's. The paradox may be explained by the continual production of helium in the form of beta particles through

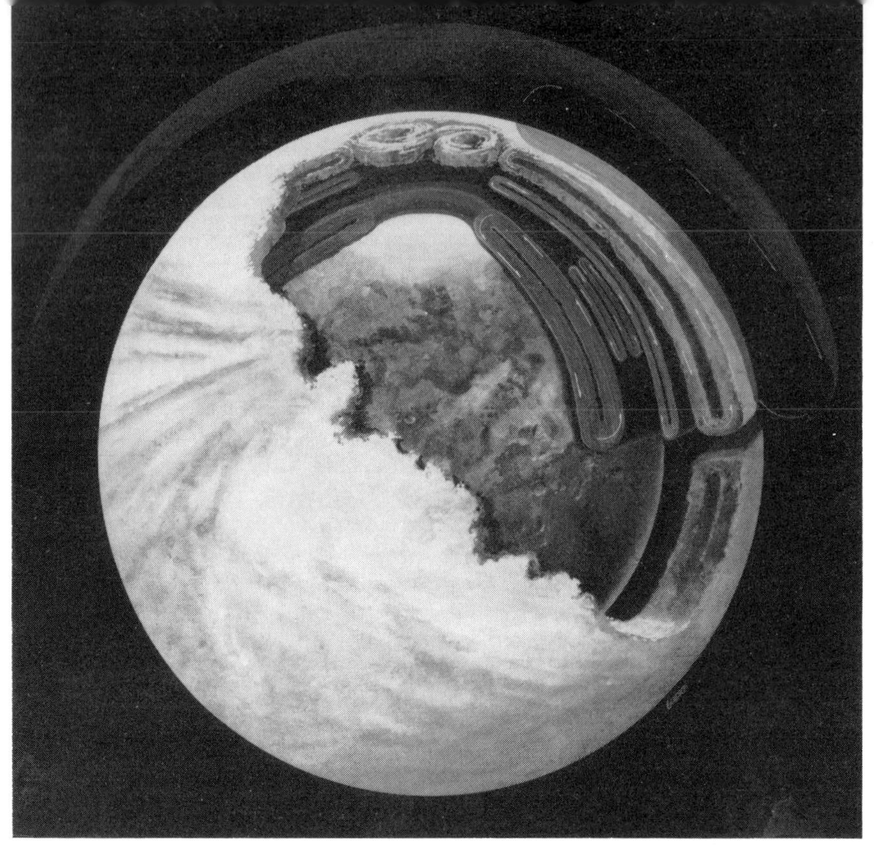

One of the major discoveries of Pioneer Venus *was the first understanding of the overall behaviour of the Venusian atmosphere, shown in these diagrams.*

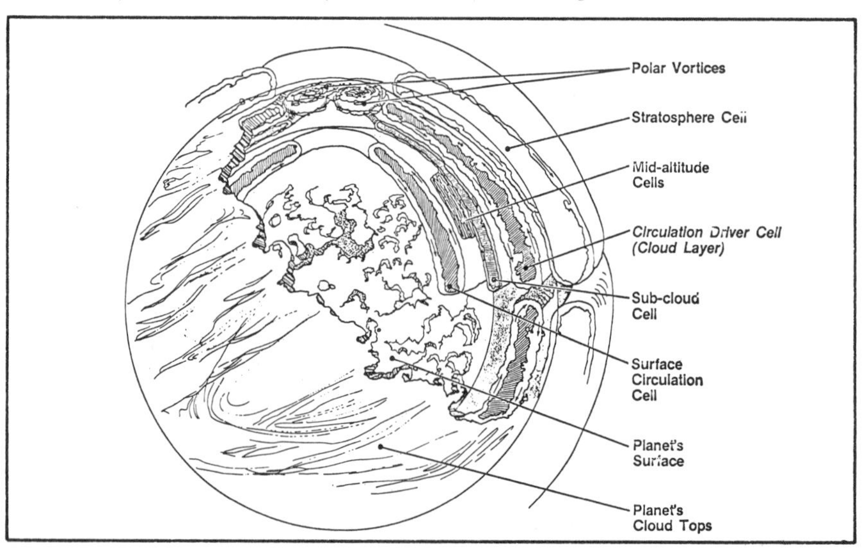

the radioactive decay of uranium and thorium in the rocks. No magnetic field was anticipated, either, but the Mariner's magnetometer did pick one up—only a hundredth as strong as the Earth's, but sufficient nevertheless to produce a magnetosphere which had a definite relationship with the solar wind.

Mariner 10 is still in the same orbit round the Sun and makes its regular visits to the vicinity of Mercury every 176 days. Its instruments and cameras are dead now and it will never resume its exploration of the smallest planet. But it has done enough to suggest answers to many of the questions which human beings have asked for thousands of years, since mankind first recognized the planet's existence.

For what has turned out to be their last mission to Venus, at least until the 1990s, NASA chose a novel method of obtaining the greatest amount of data from the minimum expenditure. The fierce environment which the Venusian atmosphere represents both in terms of pressure and of temperature makes it unlikely that any man-made artefact will survive for long. NASA therefore decided not to bother to try to land surviving craft on the planet but sought to send a cluster of probes to study the atmosphere, not expecting any of them to survive a hard landing on the planetary surface. As it turned out, one of them did, but this was an unexpected bonus from the mission which was called *Pioneer Venus.*

The dual mission began at Cape Canaveral on 20 May 1978, when the first of two spacecraft was launched by an Atlas-Centaur on a direct trajectory to Venus. This was the *Pioneer Venus* orbiter, which duly went into orbit around the planet after a flawless flight on 4 December 1978. By Soviet standards it was quite a small orbiter, with a mass of only 580 kg before orbital insertion and 370 kg after it. The Centaur injected this small mass on its planetary path in an orbit of the second type, in which the spacecraft completes more than half of a revolution round the Sun to rendezvous with the planet.

This flight of 197 days contrasts strongly with the much faster course adopted by the two Soviet spacecraft which were deployed during a previous launch opportunity, *Venera 9* and *10*. These 3 tonne monsters, launched by Proton in 1975, took only 136 and 133 days respectively to fly the short route to an orbit around Venus. The results they achieved have been discussed in an earlier chapter, but it must be said that they were not as considerable as *Pioneer*'s.

The next phase of the *Pioneer Venus* mission began more than eleven weeks after the launch of the orbiter, when a spacecraft known as the 'multiprobe bus' was fired into space on another Atlas-Centaur on 8 August 1978. This was based on a technique never before tried by the US—to aim a whole cluster of probes to a direct entry into the atmosphere of Venus. The bus was itself an instrumented spacecraft which carried four probes to the vicinity of Venus and directed them all on to carefully chosen sites on the planetary surface. The method had the merit of simplicity. Apart from small rockets for course corrections the bus needed no propulsion module, nor did the probes. All the impetus needed to direct the probes into the planet's atmosphere was supplied by the powered flight of the Atlas-Centaur, although the probes all had to be released at precise moments in the bus's spin cycle to make sure that they followed accurate trajectories to their target points. The *Pioneer Venus* orbiter was already in position when the 'multiprobe bus' and its flock of 'children' arrived just outside the atmosphere on 9 December 1978, after a quick flight of just 109 days.

Everything happened very quickly after this. The first member of the flotilla to arrive was the large probe, a 316 kg spherical instrument package which had been released from the bus on 16 November. It plunged into the atmosphere at its interplanetary speed at 18:46 hours GMT on 9 December, and took 57 minutes to descend through the atmosphere to the surface of the planet. The probe swung under its main parachute from 68 km to 47 km when the chute was jettisoned, and then, protected by its heatshield from the searing heat of re-entry, the 1.5 m sphere landed on the equator.

Its final resting place was near 43°West, close to the target point. It then ceased to transmit, which is not surprising, since the peak deceleration recorded was between 500 and 600 times the force of gravity at 78 km and the actual landing must have been hard. Its telemetry from its descent through the atmosphere was picked up by the Deep Space Network on Earth. The three small probes which followed the large probe in close formation had no parachutes and fell freely to the surface, their speed of approach being diminished only by aerodynamic braking.

Only three minutes behind the large probe came a small probe designed to land in the northern hemisphere, which had been released from the bus on 20 November. This 80 cm, 93 kg sphere landed on Venus near its target point of 75°N, 20°E at 19:46 GMT. It transmitted to the orbiter throughout its descent and exceeded all expectations by continuing to transmit telemetry for no less than 67.37 minutes after it reached the surface—a tribute to its rugged construction.

Two similar spheres, also released from the bus on 20 November, were close behind. *Small probe 2*, intended for the day side of Venus, landed at 19:51 GMT near 26°S, 45°W. *Small probe 3*, for the night side, landed four minutes later at 27°S, 45°E. Both ceased to transmit when they hit the surface. The probes had been targeted one after another to enable the Deep Space Network to pick up their signals in sequence. The telemetry showed that the probes began to be braked by the atmosphere at a height well over 100 km above the planetary surface.

The instruments in the probes were designed to study the composition and structure of the ionosphere, the atmosphere and the clouds. One interesting aspect of their descent was that the instruments' external sensors all failed at an altitude of 12 km. It is not known why this happened, but it may have been because of some electrical discharge phenomenon in the atmosphere. But there was more data to be returned from the atmosphere, for the bus, now bereft of its 'children' was also on a collision course with

Above *The* Viking *lander.*

Right Viking 1 *is launched.*

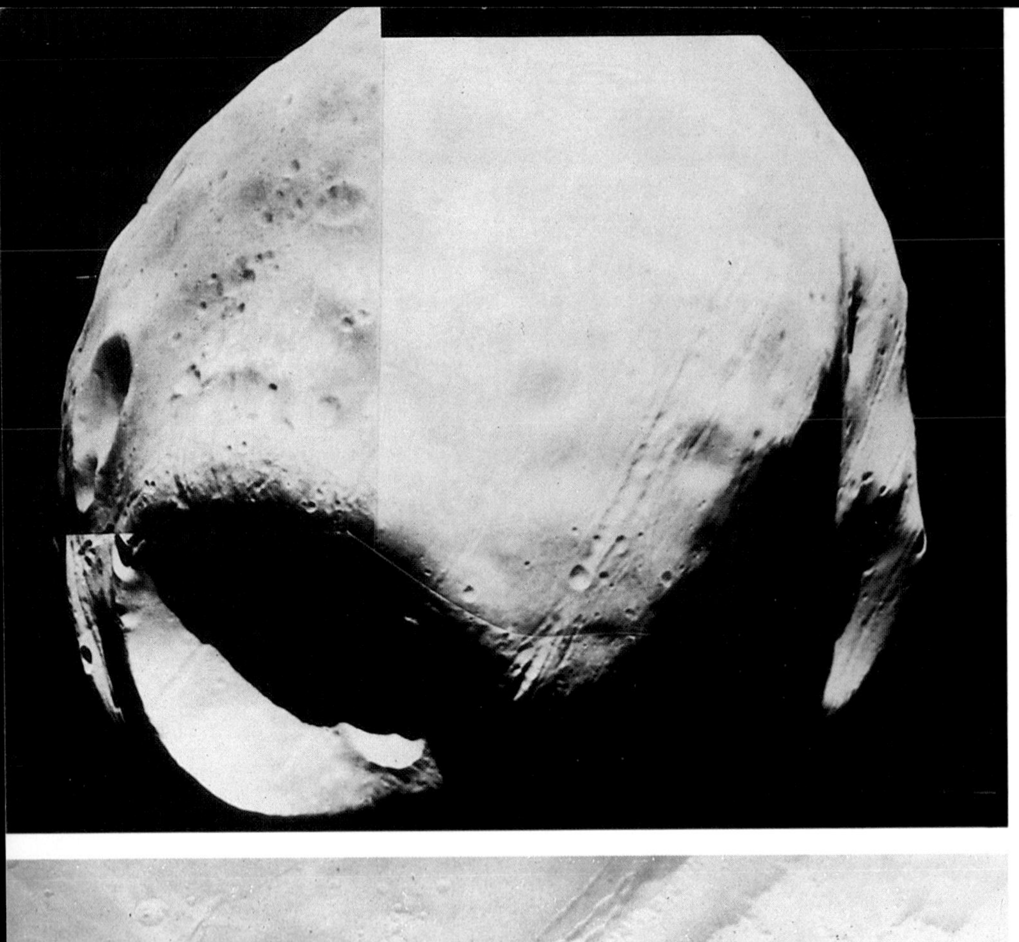

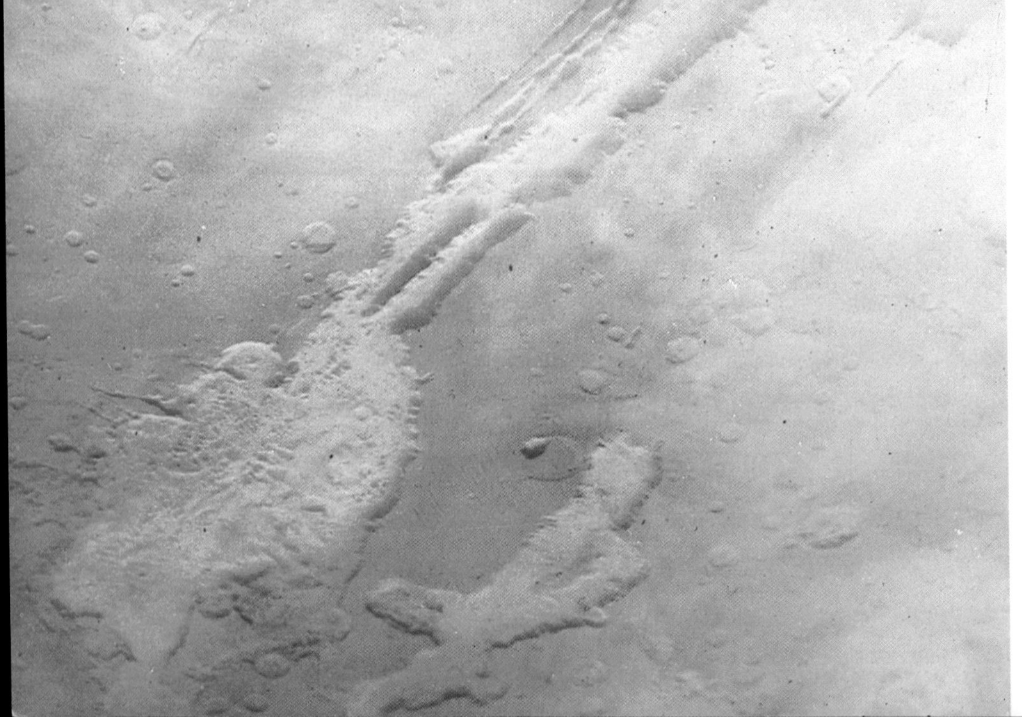

Left *Close-up of the Martian moon, Phobos.*

Below left *Valles Marineris, discovered by* Mariner, *pictured by* Viking.

Above *First picture of the Martian surface as seen by* Viking 1.

Below *Soil sampler with the completed trench in the lee of* Viking 1.

Above *Olympus Mons, the great Martian volcano.*

Below *The* Voyager *spacecraft.*

Right *Titan-Centaur rocket launches* Voyager 1.

Below *A composite picture of Jupiter and the four Galilean moons.*

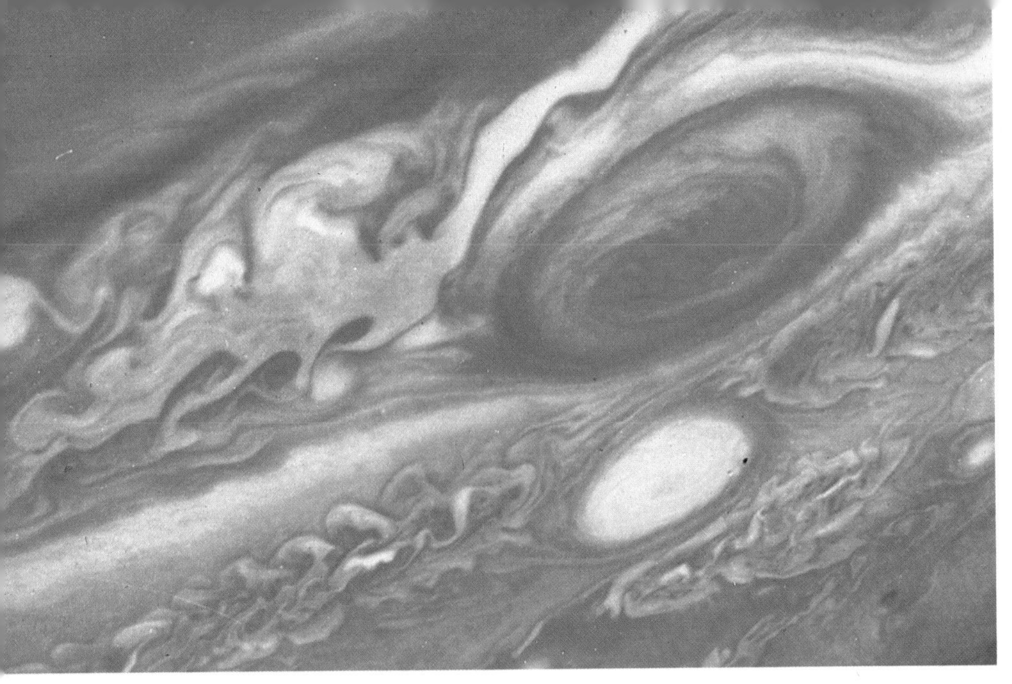

Above *The Great Red Spot of Jupiter.*

Below *Jupiter's ring scatters sunlight.*

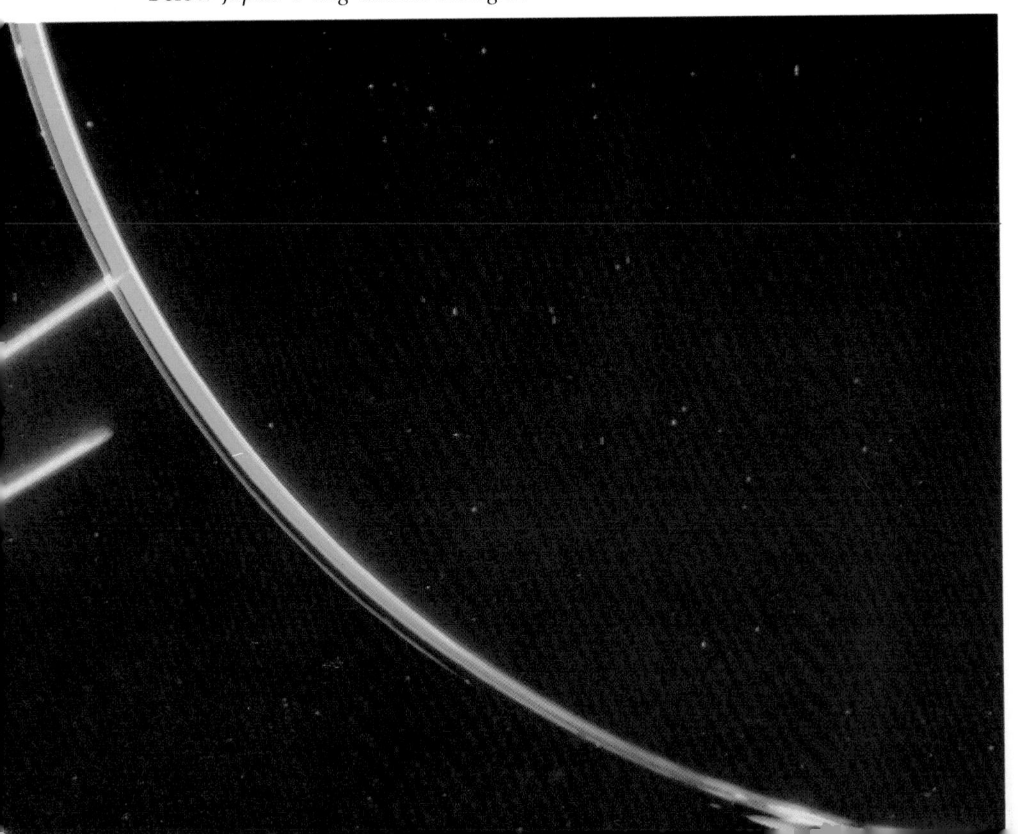

Above *Io with a volcanic plume on the horizon.*

Below *Callisto's pock-marked surface.*

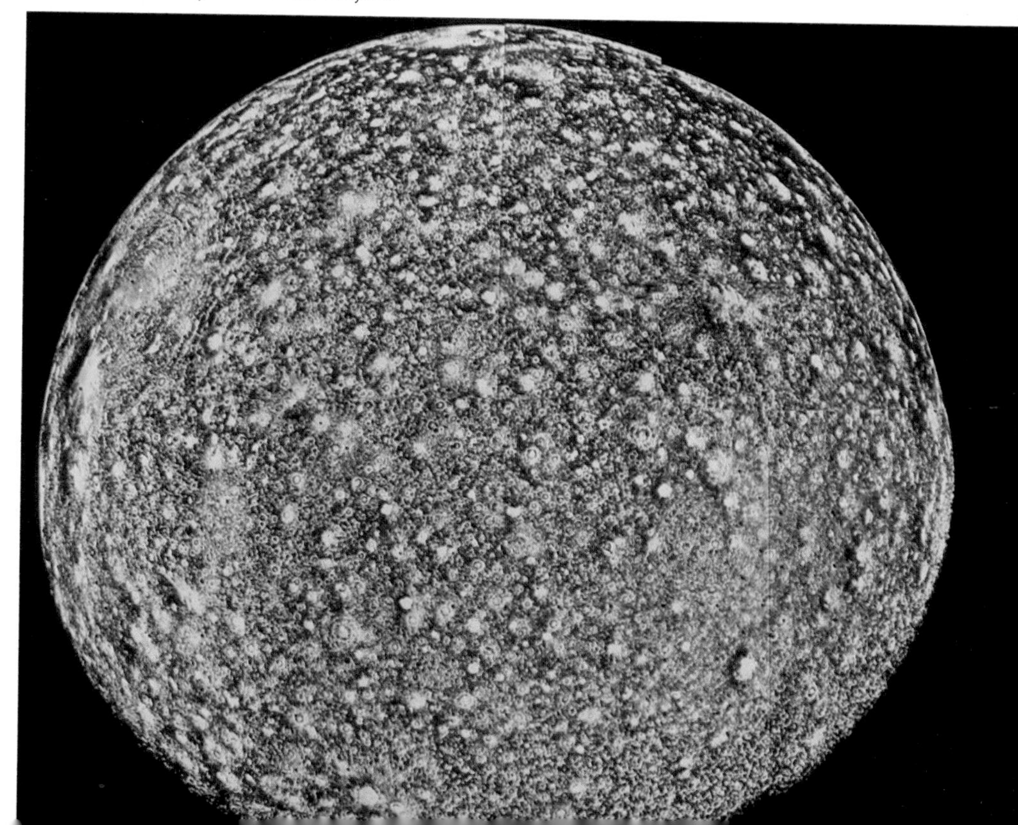

Above *Jupiter's complex clouds in close-up.*

Below *Ganymede, Jupiter's biggest moon, from 242,000 km.*

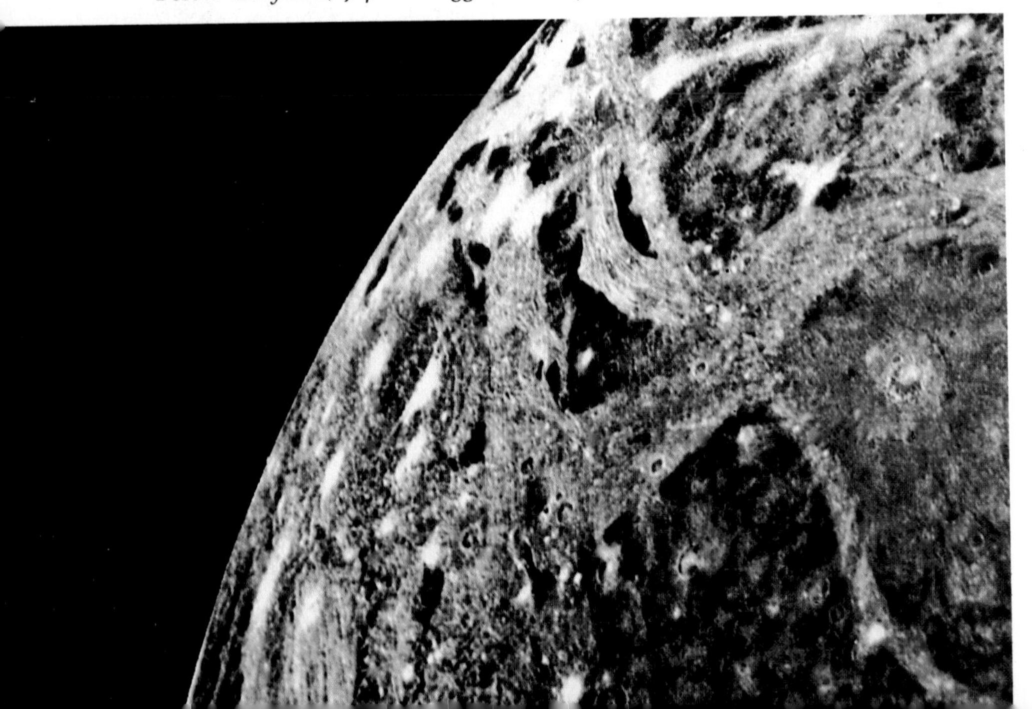

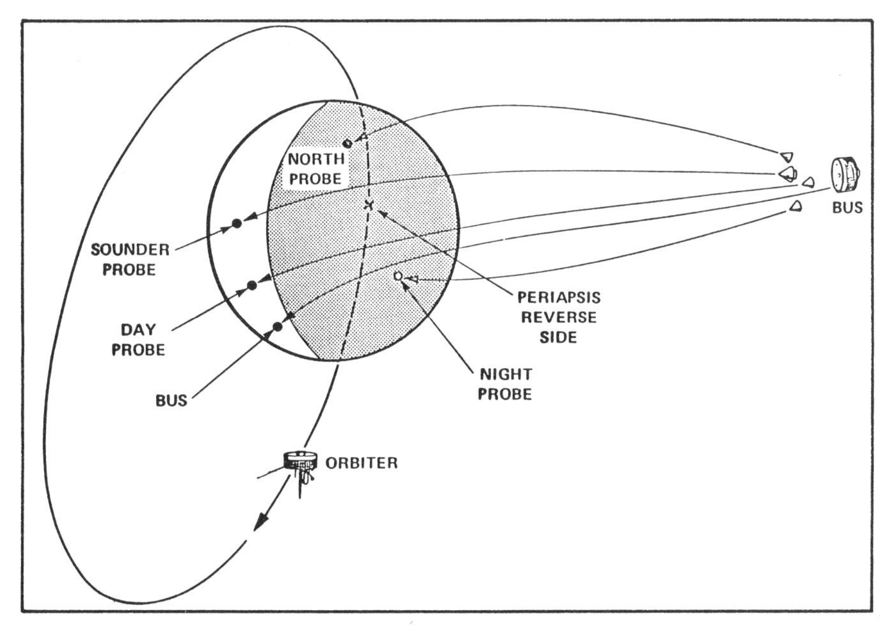

Above *The trajectories of the* Pioneer Venus Orbiter *and* Multi-probe Bus.

Below *An artist's impression of the* Pioneer Venus Orbiter. *Venus shows the cloud pattern first photographed by* Mariner 10.

An artist's impression of six components of the Pioneer *mission approaching the planet.*

Venus. It was equipped with a neutral mass spectrometer which made possible some measurements of the gases in the atmosphere before it burned up over the southern hemisphere half an hour after the last probe had landed.

The list of instruments carried by the probes is an impressive one. The large probe carried a neutral mass spectrometer and a gas chromatograph to sample the gases in the atmosphere. It also had an instrument to record atmospheric structure, a radiometer to measure the sun's net flux and an optical instrument called a nephelometer which observed the optical properties of clouds. There was also a spectrometer which measured the sizes of particles in the clouds. The small probes were equipped with nephelometers, net flux radiometers and atmospheric structure experiments. Meanwhile, the orbiter had been sent to Venus with a whole armoury of instruments which continued to study the planet for several years and was still able to observe Halley's Comet in 1986.

The orbiter was particularly useful because it was equipped with manoeuvring jets which enabled its orbit around Venus to be altered to a considerable extent. The base orbit was cleverly chosen to enable the orbiter to carry out several of the mission objectives. At first it had a periapsis of 378 km, but this was reduced to 180 km by the time the multiprobes arrived. Later it was possible to reduce it to as low as 142 km at times; positive control of the periapsis was necessary because the spacecraft was subjected to variable perturbations caused by the strong gravitational attraction of the Sun. Apoapsis was at about 65,000 km, so the orbiter took about 24 hours to complete one revolution of Venus. The 105° inclination to the equator of Venus enabled the orbiter to study much of the surface of the planet with its radar and the whole of the atmosphere with its other instruments.

The normal mission was of 243-day duration, which is the period of rotation of Venus, and during that time the periapsis travels over the planet, making possible close studies of both the day side and the night side, as well as crossing both the morning and evening terminators. This gives the instruments access to these important boundaries in the planet's 'daily' life. Since the surface of Venus is perpetually shrouded by its many layers of clouds, in some ways the most interesting task carried out by the *Pioneer Venus* orbiter was its mapping of the surface by radar. This radar did not have the power of the much larger sets carried by later Soviet *Venera* craft, but it did sterling work both in providing images of the surface and in determining how close the planet is to a perfect sphere. It achieved this by acting as a radar altimeter. In addition, the orbiter carried eleven other instruments crammed into a cylinder only 2.4 m long and 1.2 m wide. They are as follows:

ONMS—Orbiter neutral mass spectrometer;
OIR—Orbiter infra-red radiometer;
OCPP—Orbiter cloud photopolarimeter;
OUVS—Orbiter ultra-violet spectrometer;
OIMS—Orbiter ion mass spectrometer;
OPA—Orbiter plasma analyser;
OETP—Orbiter electron temperature probe;
ORPA—Orbiter retarding potential analyser;
OEFD—Orbiter electric field detector;
OMAG—Orbiter magnetometer;
OGBD—Orbiter gamma burst detector.

It was an armoury of instruments which was able during the next nine months to unravel many of the secrets of Venus and generate many miles of data and many hundreds of pages of reports. One of the most interesting of the discoveries was that droplets of sulpphuric acid are present in large quantities in the Venusian atmosphere. Since the Soviet probes have recorded lightning, perhaps associated with volcanic eruptions, the sulphur which produces the acid may very well originate in these volcanoes, as it does on Earth.

It is not true to say that there are no mysteries left on Venus for there are decades of studies to be undertaken yet before even the basic parameters of the planet are understood. But certainly *Pioneer Venus*, together with the increasingly sophisticated *Venera* probes launched by the Soviet Union in the seventies and eighties, has given the scientists who study the planets a plethora of data which will be a firm basis for future studies. The true composition and nature of the rocky surface will not be fully comprehended until a lander capable of surviving for at least a few weeks is devised, but the atmosphere is certainly now better known if not yet understood.

Pioneer Venus was able to establish new facts about the environment of Venus and its relationship with the solar wind. Although there is an interaction it does not appear to be the same as Earth's, where it is known that the magnetosphere reacts with the solar wind, heating it and causing a bow shock. There is an interaction between Venus and the solar wind, but it appears that it is the ionosphere which causes the bow wave. The planet has an extremely weak magnetic field, certainly no greater than one thousandth of the Earth's. The electric field detector in the orbiter showed that the interaction with the ionosphere is both strong and variable. The ion mass spectrometer was able to identify ions of oxygen, nitrogen and carbon chemistry in the ionosphere. This study was backed up by the retarding potential analyser, which also recorded the thermal structure of the ionosphere.

The OCPP obtained images of Venus in the ultra-violet for studies of large scale cloud morphology, cloud-tracked winds, and the size and shape of cloud particles. The probes established that there were variations in the pressure of the extraordinarily dense Venusian atmosphere at the surface. This variation, from 86.2 to 94.5 times Earth's surface atmospheric pressure, is probably explained by the different altitudes of the sites where the probes came to rest. There was a similar variation in temperatures, from 721°to 732'K, for the same reason.

One interesting result from the nephelometer on the small probe which survived its hard landing was evidence that a dust cloud rose from the surface on impact, and then subsided. Perhaps the probe survived its collision with the planet because it happened to land in an area of deep dust.

The solar flux radiometer, measuring the amount of sunlight getting through, showed that there were three cloud layers above an altitude of

US missions to Venus

Name	Date	Result
Mariner 1	25 July 1962	Guidance failure during launch
Mariner 2	27 August 1962	Success: 40,000 km fly-by
Mariner 5	14 June 1967	Success: close fly-by
Mariner 10	2 November 1973	Success: close fly-by + Mercury
Pioneer Venus	20 May 1978	Success: Venusian orbit
Multi-probe	8 August 1978	Success: four probes deployed

49 km. Below that height the atmosphere was clear and the visibility varied between two and eight kilometres. Finally, only two per cent of the sunlight falling on the cloud layer reached the planetary surface.

Previous studies had shown that the atmosphere of Venus is largely composed of carbon dioxide and *Pioneer Venus* confirmed this, giving a CO_2 proportion of 96.4 per cent. This means that there is just about as much carbon dioxide on Venus as there is on Earth. The difference is that on Earth most of the gas is tied up as carbonate compounds in rocks like limestone because we have a highly oxidizing atmosphere and there is very little in the atmosphere. On Venus, where the ambient temperature is so much higher, the carbon dioxide has been driven out of the rocks and is mainly atmospheric. Nitrogen accounts for 3.41 per cent of the atmosphere, water vapour for 0.13 per cent and there are traces of oxygen, argon, neon and sulphur dioxide. In addition to these gases there are major quantities of sulphuric acid in aerosol drops in the clouds and sulphur compounds in the lower atmosphere suggest that there is a form of acid rain of a highly corrosive nature that does not actually reach the ground.

In 1973, *Mariner 10* took pictures of Venus for eight days at resolutions from 130 km to 100 m. They established that the atmospheric flow is symmetrical between the northern and southern hemispheres and that the winds at the equator reach 100 mps (350 kmh) at a high altitude. One unexpected finding was that immediately beneath the Sun on the equator of Venus a disturbance continually develops, within which a cellular structure reminiscent of convection is seen. The equatorial region has a mottled appearance in some areas, while streaks and whorls are seen at high latitudes. Before the flight, in fact since 1926, a horizontal dark 'Y' marking has been known from telescopic observation in the ultra-violet. *Mariner 10* took superb photographs of this feature which now represents the visual appearance of Venus as it is best known, now that we no longer rely on ground-based astronomy. When *Mariner 9* went to Mars its communications bandwidth made it possible to transmit at 16 kilobits a second. For *Mariner 10* this figure was 117.6 kbs, so it was not necessary to record pictures on tape for later transmission. The pictures were made up of pixels (picture elements) each of which could be one of 256 discrete levels, ensuring high resolution of details on the planetary surface.

At this point it is possible to sum up the main features which have been discovered as a result of the many missions to Venus by both America and the Soviet Union. Radar studies, both from big radio-telescopes on Earth and from the space vehicles such as the *Pioneer Venus* orbiter and *Venera 15* and *16*, have begun to fathom out the geomorphology and the figure of Venus. The figure of a planet describes how far it deviates from a perfect sphere. In the case of the Earth there is oblateness, that is, a flattening at the poles, and also an equatorial bulge. Venus, it has been discovered, has neither and no less than 20 per cent of its surface lies within 125 m of its mean radius of 6,051.4 km.

A remarkable 90 per cent of the planet lies within 3.5 km of this mean radius, which makes it very close to a perfect sphere. Even so, there are some mountains and the highest are in the area in the northern hemisphere known as Ishtar Terra and first identified from Earth. This is a 'continent' as big as Australia standing as high as 11.1 km above the surrounding plain. Ishtar Terra contains the highest mountains on Venus, known as Maxwell Montes and other ranges which have been given the names Lakshmi Planum, Akna Montes and Freijas Montes.

Another similar area, Aphrodite Terra, is as big as Africa, but has a lower elevation of 5 km above the mean level. Together with a third region, Beta Regio, these areas may be lava-covered uplifted segments of low density crust surmounted by large volcanic peaks. At the top of Maxwell Montes radar has pinpointed a caldera 100 km across and 1 km deep, and two peaks on Beta Regio, called Theia Mons and Rhea Mons, are believed to be active volcanoes. The Beta Regio rocks were identified as alkaline basalt by *Venera 13* and *14*. Apart from the obvious differences of atmosphere and oceans, the contrast with the Earth could hardly be greater, although Venus is virtually a twin of our planet in size and density. Earth's continents and oceans have been formed by the process known as plate tectonics (once known as Continental drift). The continents are situated on plates of thin crust which drift over the surface of the globe propelled by convection currents in the mantle. Venus has no such history. Its crust seems to be all one plate which is considerably thicker than the Earth's. We have volcanoes in many parts of the world where plates clash together, but on Venus volcanic activity is strictly local.

Space probes have established that the ionosphere of Venus extends to an altitude of 400 km and the top of the main atmosphere is at 200 km, where the temperature is 27°C, although 90 per cent of the atmosphere lies below an altitude of 28 km. At 70 km the first clouds are encountered and a dim Sun can still be seen at 66 km, but by 63 km the Sun is diffused. At this level the temperature is down to 13°C, the pressure is about half that of Earth at sea level, and visibility is down to 6 km. At 50 km visibility is less than 2 km, the temperature is 20°C, and then there is a clear layer from 47 km where the temperature is 200°C. In this layer a haze of particles of about one micron in size and at a concentration of between 1 and 20 per cubic centimetre was found, but below 38 km the atmosphere is particle-free. Visibility is as great as 80 km and the temperature is 310°C at 30 km, and 380°C at 20 km. There is a reddish outside light which colours the sky and the surface of the planet becomes visible at an altitude of 7 km. At the surface visibility is 3 km.

The 'greenhouse effect' has long been known to dominate in the atmosphere of Venus but the surface temperature of 737°K (that is, 737° above absolute zero, or about 463° Celsius) was something of a surprise. The effect is caused by the inability of the longer wavelength infra-red rays reflected from the planet to penetrate various gases although they are transparent to the incoming sunlight. This effect is currently a worry on Earth, since the use of fossil fuels is increasing the slight amount of carbon dioxide in our atmosphere, which it is feared will heat up our atmosphere with potentially catastrophic results.

On Venus, although it represents more than 96 per cent of the atmosphere, carbon dioxide is responsible for only 55 per cent of the greenhouse effect. Water vapour, which is present at very low concentrations, adds a further 25 per cent, and sulphur dioxide, making up only 0.02 per cent of the atmosphere, is responsible for 5 per cent of the greenhouse effect. The remaining 15 per cent is due to the clouds and hazes which are ever-present.

Chapter 6
The red planet

Of all the planets, Mars has aroused more curiosity and anticipation among men than any other in the Solar System. Even before the Italian astronomer, Schiaparelli, claimed to see 'canali', or channels on the surface of the 'red planet', scientists had speculated that it was here, if anywhere, that we would encounter sentient beings. 'Martians' were synonymous with aliens who would either expand man's horizons by bringing him new and exotic knowledge, or would be the nemesis for our species by bringing death and destruction.

With this background, it is not surprising that some of the most exciting and fruitful planetary missions have been aimed at Mars. What is curious is that the Soviet Union, with its long and sustained campaign of research applied to Venus, has been half-hearted in its approach to Mars. True, its attempts to explore Mars spanned a total of thirteen years, from 1960 to 1973, with more than a dozen craft involved, but this is nowhere near the magnitude or the intensity of the *Venera* programme.

Why this should be is a mystery, for Mars is just as interesting a target as Venus and the difficulty of reaching it is no greater. Yet the Soviet Mars missions ceased in 1973 and at the time of writing have not been resumed. It is possible that budgetary pressures are responsible, but there have been many rumours that the Soviet Union intended in the not too distant future to send a manned expedition to Mars and it would have been logical to have carried out a thorough reconnaissance of the planet in advance of such a risky venture.

Launch windows for Mars come at approximately 25-month intervals and it was in the 1960 window that the first tentative steps were taken by the Soviet Union. On 10 and 14 October, the United States revealed later, there were attempts to place payloads into Earth orbit as the first step in sending instruments to Mars. Both failed even to enter Earth orbit, which was a bitter disappointment for the Soviet leader, Nikita Kruschev, who was attending a meeting of the United Nations General Assembly in New York at the time and apparently had with him on the ship in which he had sailed a facsimile of the Mars spacecraft. The propaganda coup failed and the model went back to the Soviet Union without being displayed to a wondering world.

The next launch window was in 1962 and in this year the Soviet Union achieved at least partial success. Two out of three attempts, on 24 October and 4 November, reached Earth orbit, but the same trouble that was afflicting *Luna* and *Venera* flights at the time, the failure of the escape stage, stranded them both there and they were not even acknowledged by Moscow. However, sandwiched between them was a successful escape stage firing and the payload, christened *Mars 1*, was placed into a trajectory heading towards the 'red planet' on 1 November 1962. It was a sizeable craft, with a mass of 893.5 kg, and an improved version of the *Venera* which became the standard craft for planetary research until the introduction of the larger payload of the Proton rocket with *Luna 15* in 1969 and *Venera 9* in 1975. However, *Mars 1* was not a success in research terms, for although it is estimated to have passed within 193,000 km of Mars in June 1963, radio communications with it ceased on 21 March and therefore no research findings were received.

The next window was at the end of November 1964, and a Mars probe was duly launched on the last day of that month. But it was given the name *Zond 2* rather than *Mars 2*, although Moscow did accurately announce that it was intended to study Mars. However, the unreliable radio system failed once again, in April 1965, and although *Zond 2* passed within a creditable 1,500 km of Mars on 6 August it was, like its predecessor, a complete failure in scientific terms. There were probably other attempts to launch payloads during this window but they were not acknowledged and they did not reach Earth orbit. The next Mars-related launch was on 18 July 1965, when *Zond 3* was ejected from Earth orbit on a trajectory which would cross the orbit of Mars at a point not then occupied by the planet. This was obviously an engineering trouble-shooting flight, designed to iron out the radio shortcomings that were plaguing the Mars programme.

Zond 3 was equipped with cameras which were used to take 25 pictures of the Moon as it flew by our satellite at a distance of 9,200 km. The pictures were of a better quality than those that had been taken by *Luna 3* six years before and they were transmitted back to Earth repeatedly as *Zond 3* receded. The spacecraft was still transmitting when it crossed the orbit of Mars, but of course at this time it was too far away from the planet to return any scientific data or pictures. Although the flight was a success from the operational point of view there was now a long gap in Mars activity and it was with a new spacecraft, built as the payload of the Proton rocket, that the Soviet campaign to explore Mars was resumed. Some success was achieved, but it was not total.

Two launch windows had been missed, but on 10 May 1971, a payload was placed into Earth orbit at the time and with the parameters which indicated that it was a Mars craft. It became known as *Cosmos 419* when it failed to leave Earth orbit, but it was quickly followed by *Mars 2* on 19 May and *Mars 3* on 28 May. Having abandoned the smaller, 990 kg type of craft, the Soviet planners had been able to build in each case a combined orbiter and lander,

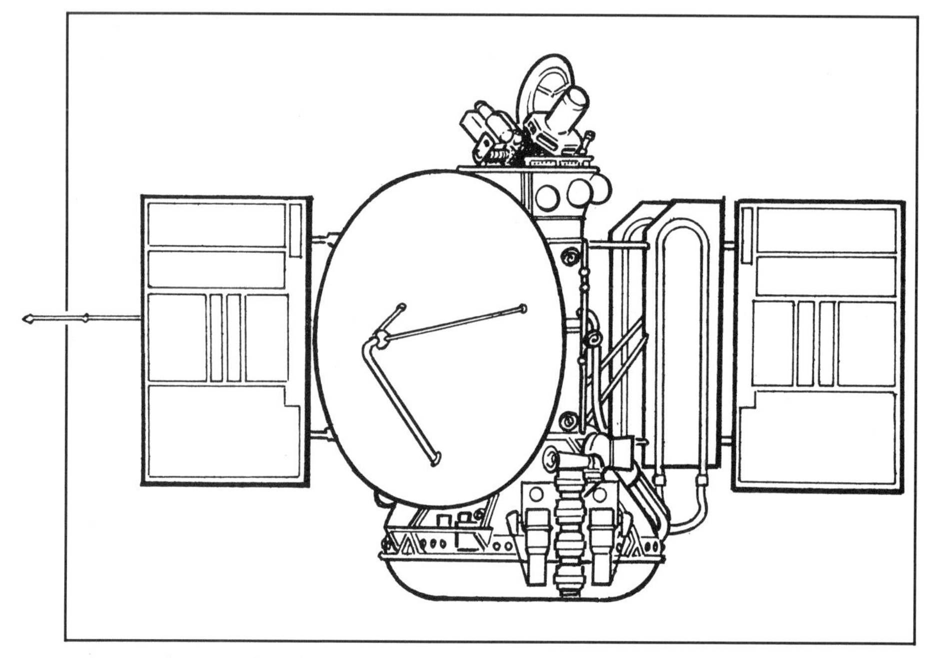

Mars 4 and 5 were the orbiters taking part in the 1973 mission to the red planet. Their task was to relay back data from the landers.

with a total mass before injection into Martian orbit of about 4,600 kg. After the insertion motor had burned its fuel the orbiter had a mass of just over 2,000 kg and the lander, 1,200 kg in Martian orbit, was down to 635 kg once it had descended beneath its parachute to the Martian surface.

With *Mars 2* it was a case of so near and yet so far. Everything appeared to work normally until the very last moment. Martian orbit was achieved on 27 November 1971, the lander was detached, its retro-rocket was fired and the descent began. Although the impact point in the southern hemisphere is known, it appears that no signals were received from the lander after it touched down, so presumably it was damaged or destroyed on impact.

Mars 3, which was a twin of *Mars 2*, had better luck. It was injected into orbit round Mars on 2 December and the same day it ejected its lander which made a successful soft landing on another part of the southern hemisphere. This lander did survive the landing but not for long; the orbiter, which was acting as a relay for the weak signals from the surface, broadcast data for a mere twenty seconds, and it is thought that the signals from the lander may have been cut short by the effects of a dust storm that was raging at the time.

However, the orbiters were equipped with a full range of instruments as well as two cameras with wide angle and telephoto lenses. They continued to operate from their eccentric orbits round Mars, with an inclination of 48.9°, for about four months and returned some significant data as well as pictures. The camera system used was similar in principle to that used long before in *Luna 3*. Ordinary photographic emulsion was used, the pictures were developed on board and then scanned electronically, each of 1,000 lines containing 1,000 picture elements. The million data points for each picture were then transmitted to Earth by radio. Results were disappointing because the same dust storm which also interfered with *Mariner 9*'s efforts to photograph the Martian surface, obscured details.

Instruments on board the orbiters identified atomic hydrogen and oxygen in the upper atmosphere of the planet, studied surface relief by measuring the amount of carbon dioxide along a sighting line and determined water vapour concentrations. Temperatures were found to vary from -110°C to 13°C and differences of relief as great as 4 km were mapped. Scientific data returned by the two spacecraft were relatively sparse considering the complexity of the missions, but it was a good start to Martian exploration after a shaky beginning to the programme.

At the next launch window the energy requirements for reaching Mars were more stringent and although the Soviet Union's next venture was of the same order of complexity as *Mars 2* and *Mars 3*, this time no fewer than four vehicles were needed. This was becuase even the Proton rockets did not have sufficient power to place both the orbiter and the lander on a trajectory out of Earth orbit to Mars.

The orbiters, *Mars 4* and *5*, went first, on 21 and 25 July 1973, and they were followed by the landers, *Mars 6* and *7*, on 5 and 9 August. The intention was, in this large-scale and, as it turned out, last Soviet expedition to Mars to date, that the first two should go into orbit around the planet about a month before the landers arrived and should act as communications relays as the previous orbiters had. Unfortunately, this ambitious and complex operation went badly wrong. First, the retro-rocket of *Mars 4*, the first arrival, failed to fire when it reached Mars on 10 February 1974. As a result, it merely continued past the planet in a heliocentric orbit, although from its closest approach point of only about 2,200 km it was able to take pictures which were transmitted to Earth by the same system used on the earlier mission.

Next to arrive was *Mars 5*, two days later, and it did go successfully into orbit, with a period of 25 hours. The orbit was at an inclination of 35°, with an apoapsis of 32,500 km and a periapsis of 1,760 km. This, as it happened, was the only member of the quartet to operate satisfactorily, for both of the landers failed to complete their part of the operation. *Mars 7*, the first of the two to arrive, detached its lander successfully on 9 March, but there was a failure in an onboard system and the capsule missed the planet by 1,300 km.

Finally, *Mars 6* did carry out the necessary manoeuvres on 12 March and the capsule descended through the tenuous Martian atmosphere to a point in the southern hemisphere. But 2½ minutes after the parachute opened and just before touchdown contact was lost and never recovered. The expedition

Left Mariner 4's *solar panels contained 28,224 separate solar cells which provided the power for the remarkable pictures of Mars.*

Right Mars 6 *and* 7 *were almost identical to* Mars 4 *and* 5, *but with the addition of landers intended to go down to the Martian surface* (D. R. Woods).

Below right *The growing complexity of interplanetary craft is demonstrated by this picture of* Mariner 4's *television camera.*

was a disaster which was retrieved only slightly by the excellent research work carried out by *Mars 5* in orbit.

Apart from some readings of chemical composition and temperature and pressure of the atmosphere from *Mars 6*, the only significant scientific data came from *Mars 5*. It was equipped with a formidable battery of instruments to study all aspects of the planet and its environment and also two television cameras, replacing the conventional photographic system of the earlier probes. There were some interesting variations from the data returned two years earlier by *Mars 3*. At that time the dust storm that was raging planet-wide must have dehydrated the atmosphere, for *Mars 3* found only minute traces of water vapour. The readings from the instruments on *Mars 5* revealed a level four to eight times as great.

As an example of international co-operation, this Soviet campaign was a model. The photographs taken by the *Mars 5* cameras were studied by both American and Russian planetologists and it was possible to match them with pictures from *Mariner 9* two years before and so provide control points from which proper maps could be drawn up.

Having established this basis of research into Mars and its environment the Soviet Union then ceased to follow up the subject. But the American interest in Mars has been more wide-ranging and also rather more sophisticated. It began with the failure of *Mariner 3* in 1964 but went on to great success and culminated in the intriguing biological study of the Martian soil in the *Viking* programme of 1975-76.

The *Mariner* Mars mission in 1964 was intended to be a twin flight of Atlas-

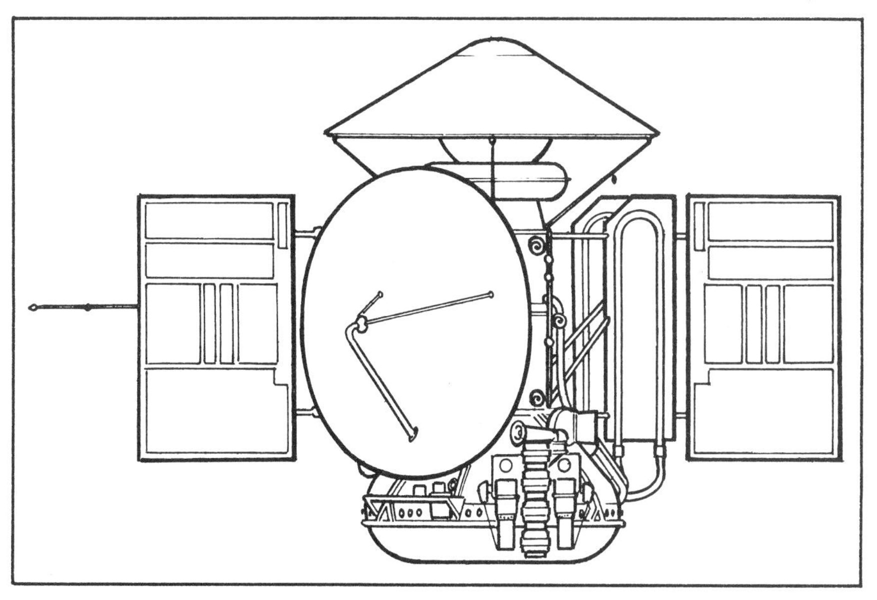

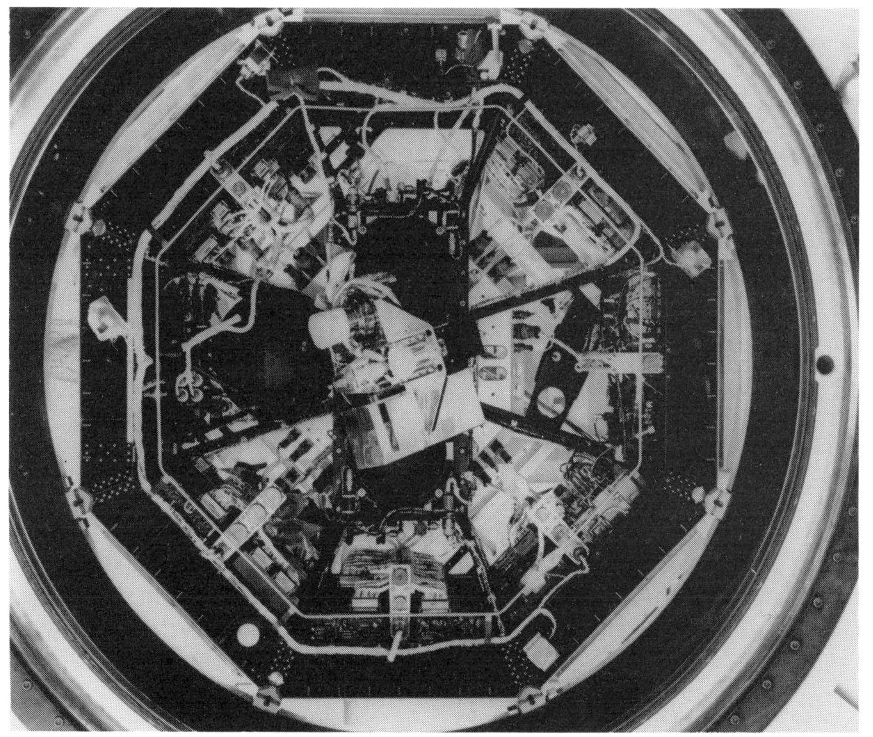

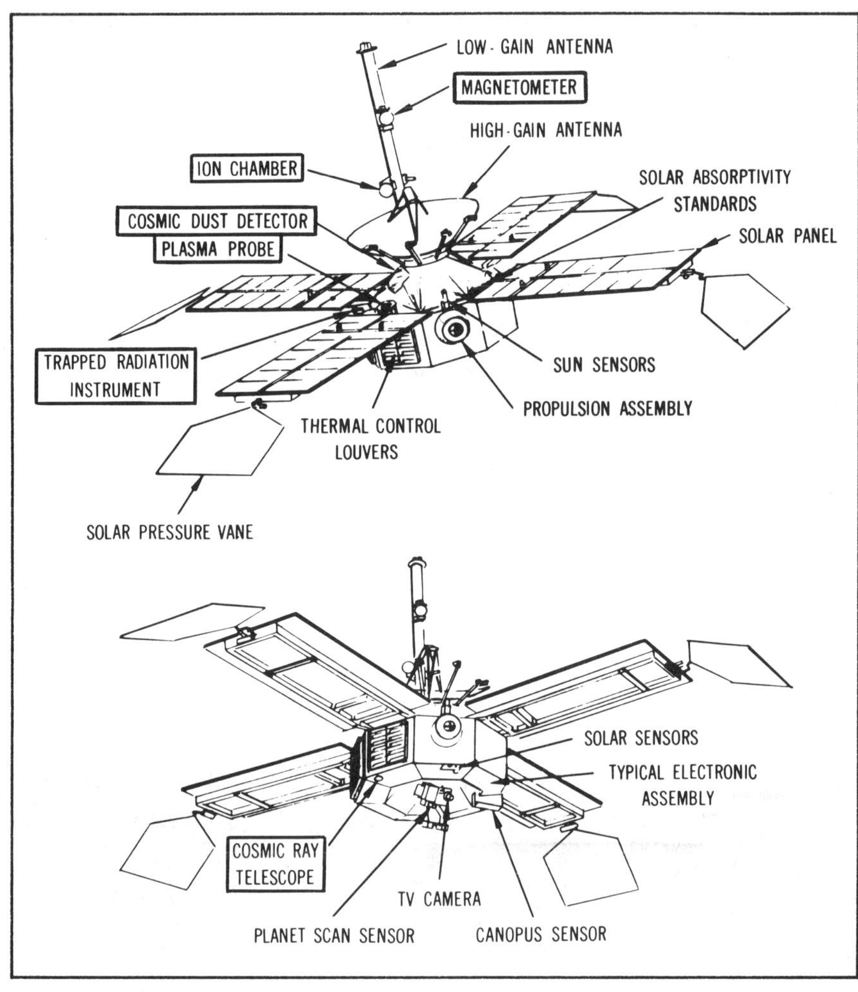

The component parts of Mariner 4.

Centaur payloads each of several hundred kilogrammes, but as in other programmes the delays in the Centaur development meant that a much smaller craft had to be substituted. The Atlas-Agena which had to be used as the launch vehicle could inject 261 kg into the Martian trajectory in the favourable 1964 geometry and that was the weight of the *Mariner 3* and *4*. They were based on the *Ranger* spacecraft which had finally succeeded in lunar research (see Chapter 3), but they were much more complex, with over 130,000 parts in each spacecraft.

Compared with *Mariner 2*, the new generation was upgraded with four solar panels instead of two (necessary because the trajectory took it further from the Sun) and a television camera, since a primary aim of the two missions was photography of the Martian surface, seen up to then only by means of telescopes and through the murky Earth atmosphere. It was not an advanced camera by later standards; it was planned to take only 21 pictures and store them in digitized form, to be transmitted later at the abysmal rate of one picture every 8 hours 20 minutes. But it was the first to attempt the feat of picturing the surface of Mars and the first to succeed.

Not the first time, however. *Mariner 3* was launched on 5 November 1964, and at first seemed to be going well. The Atlas burn was nominal, but the shroud that covered the payload failed to separate to order. Although the Agena fired and injected the *Mariner* into a course towards Mars the mission failed because the spacecraft could not work with its solar panels and antennas covered by the shroud.

A quick re-design of the shroud enabled a more successful mission to begin only 23 days later, when *Mariner 4* began its epic journey to the Red Planet. This time there was no mistake and although the Atlas-Agena placed the probe into a trajectory which would have missed by 240,000 km, it was possible to correct the course by a brief burn of the hydrazine-powered engine on 5 December. During its cruise phase *Mariner 4* was able to measure parameters of the interplanetary medium, including cosmic rays, solar radiation, meteoroids, magnetic fields and particles.

Mariner 6 *and its twin,* Mariner 7, *had ten times the photographic resolution of* Mariner 4.

Left *The small dimensions of the* Mariner 7 *compared with later spacecraft can be seen in this view of it being placed in its protective launch shroud at Cape Canaveral.*

Right *The scan platform hanging below the* Mariner 9 *spacecraft carries the cameras which transmitted thousands of pictures of Mars.*

Closest approach was on 15 July 1965, with an acceptable miss distance of 9,851 km after a voyage of 230 days. The 30 cm focal length Cassegrain telescope of the television system had been designed to cope with a range of illumination of 30 to 1, ranging from full solar illumination to near total darkness at the terminator. It duly took 22 pictures of the Martian surface at ranges which varied from 17,000 to 12,000 km. They were history-making images which revealed that the enigmatic surface of the planet was liberally sprinkled with craters—at least on the 1 per cent of the surface that the *Mariner* covered. There were seventy clearly distinguishable craters, ranging in diameter from 4 to 120 km, with rims rising about 100 m above the surrounding surface and depths of many hundreds of metres. It was a surface remarkably similar in size distribution of the craters to the uplands of the Moon, despite the differences in size and environment between our satellite and Mars. *Mariner 4* was able to put an upper limit on the Martian magnetic field of not more than one thousandth of the terrestrial field.

There was therefore no surprise in the discovery that there is no radiation belt around Mars because a much stronger magnetic field is needed to trap the energetic particles which make up a belt such as that which surrounds the Earth. But *Mariner* did find that interplanetary dust is more abundant in the vicinity of Mars than it is near the Earth. It also established that the density of the atmosphere was only about 1 per cent of the Earth's. With a data transmission rate of only 8.33 bps *Mariner 4* was still able to return to Earth a healthy package of data and images that radically transformed the conventional view of Mars as somehow comparable with the Earth.

America, like the Soviet Union, missed the next launch window but in 1969 had, at last, the opportunity to use the Atlas-Centaur payload to send more capable *Mariners* to Mars. JPL was able to design a craft with a weight

of 413 kg and it was this payload that was launched twice, as *Mariner 6* and *Mariner 7*, on 25 February and 27 March. This latest version of the Mariner was of the same configuration as its predecessor but far more instruments could be included. To begin with, the single fairly simple camera of *Mariner 4* was replaced with two TV cameras which were placed on a movable scan platform. While one camera took wide angle pictures the other provided telephoto images and there were two tape recorders associated with them. One was a digital tape recorder which was used to store the six least significant bits of an eight-bit digitally coded signal from every seventh picture element of each picture line. This was the most econnomical way of sending back picture data even at the new transmission rate which was 2,000 times faster than *Mariner 4*. An analogue tape recorder was used to store the analogue video signal from each pixel, and the two most significant bits of the eight-bit word were averaged over several lines and transmitted in real time. It was a far cry from the 8.33 bps transmission rate of only four years before.

Mariner 6's trajectory in a favourable launch window was a swift one and the craft arrived in the vicinity of Mars on 31 July 1969. Its success was rather overshadowed by the fact that only a few days before, the crew of *Apollo 11* had arrived back on Earth after their successful first landing on the Moon. Nevertheless, there was an air of great anticipation at the JPL control centre as the close encounter approached. Those of us who were there will never forget the intense excitement as this new wave of planetary exploration began.

Far encounter pictures began returning from *Mariner 6* 54 hours before closest approach. Between 48 and 28 hours the narrow angle camera was used to obtain 33 full disc pictures of Mars, and they were transmitted over the next three hours. From 22 to 7 hours, seventeen more far encounter pictures were recorded, and then, during eighteen minutes of the close approach, during which *Mariner 6* came as close as 3,430 km to Mars, the two cameras took 25 pictures of the surface.

For the first time, the southern polar cap appeared in some of the pictures. It was revealed as a sharply bounded area of ice or snow, with several craters showing up along the edge. One frame covering a total of 625,000 km^2 of the surface was found to contain 156 craters varying from 3 to 240 km in diameter. In telephoto views many more craters down to 300 m in diameter were seen. It was, as *Mariner 4* had shown, rather like the lunar surface, but there were some differences. The relief of Mars is more subdued and there is an absence of obvious secondary craters and rays of ejecta.

The scan platform carried both an infra-red radiometer and ultraviolet and infra-red spectrometers and these instruments were for the first time able to produce data on the constituents of the Martian atmosphere. As expected, the tenuous atmosphere was largely carbon dioxide, but it was also found to contain ionized carbon monoxide and atomic oxygen and hydrogen,

Missions to Mars

Name	Date	Result
Unnamed	10 October 1960	Did not reach Earth orbit
Unnamed	14 October 1960	Did not reach Earth orbit
Unnamed	24 October 1962	Failed to leave Earth orbit
Mars 1	1 November 1962	Passed Mars but radio failed
Unnamed	4 November 1962	Failed to leave Earth orbit
Mariner 3	5 November 1964	Shroud failed
Mariner 4	28 November 1964	Success: 22 pictures
Zond 2	30 November 1964	Passed Mars but radio failed
Zond 3	18 July 1965	Successful engineering test
Mariner 6	25 February 1969	Success: 75 pictures
Mariner 7	27 March 1969	Success: 126 pictures
Mariner 8	8 May 1971	Centaur rocket failure
Cosmos 419	10 May 1971	Failed to leave Earth orbit
Mars 2	19 May 1971	Into Mars orbit, lander failed
Mars 3	28 May 1971	Into Mars orbit, 20 seconds of signals
Mariner 9	30 May 1971	Success: 7,000 pictures
Mars 4	21 July 1973	Failed to orbit Mars
Mars 5	25 July 1973	Into orbit
Mars 6	5 August 1973	Lander sent no signals
Mars 7	9 August 1973	Lander missed by 1,300 km
Viking 1	20 August 1975	Orbiter + lander
Viking 2	9 September 1975	Orbiter + lander

although no nitrogen was detected. A temperature as high as 16'C was measured by the radiometer on the equator at noon, but on the night side of the planet this temperature could drop to as low as -73'C.

Mariner 7 was on a slightly faster trajectory than its twin and made its closest approach to Mars on 5 August 1969. It took 93 far encounter pictures which appeared mottled and 33 at the near encounter. The mottling turned out to be caused by craters with diameters up to hundreds of kilometres. The 'canali' of Schiaparelli, never absolutely disproved by telescopic observation, were seen on the pictures to be made up of discontinuous linear features and chance associations of craters and other markings. The study of climàtology on Mars began with the *Mariner 7* pictures, for some of them exhibited suggestions of clouds—a hint that was proved up to the hilt by later missions, which detected many clouds.

The time had now come to place instrument packages in orbit around Mars—a decision that had already been made by the Soviet planners. A veritable fleet of explorers took off in May 1971.

Although *Mariner 9* was the last of the three to leave Earth, it was the first to arrive, beating *Mars 2* by thirteen days and *Mars 3* by eighteen. But the launch window was not the occasion for complete triumph for America, for *Mariner 9* was only part of a twin mission of which the other half, *Mariner 8*, failed. This time, it was the guidance system of the Centaur upper stage that did not function properly after the Atlas booster had given the mission a good start. *Mariner 8* did not reach Earth orbit but plunged into the Atlantic to its destruction on 8 May 1971. Some urgent re-programming now had to be put in hand, for the two spacecraft had different roles in the exploration of Mars.

Mariner 8 was to have been put into a polar orbit, covering the whole of the Martian surface in the course of three months and returning up to 5,400 pictures. *Mariner 9*'s orbit was to have been at an inclination of 50°, where it would be synchronized with the planet's rotation in such a way that its ground track would repeat, thus making any changes in surface features observable. In the frantic few days before the window closed, the *Mariner 9* flight plan was modified to give a 65° inclination and a ground track repeating itself every seventeen days. This was a way of maximizing the data and images returned now that the loss of *Mariner 8* made the full mission impossible.

Only three weeks after the loss of *Mariner 8*, its sister-ship took off from Cape Canaveral on 30 May 1971, and this time the Atlas-Centaur combination worked perfectly and injected to the craft towards Mars. The new explorer was based on the design for *Mariner 6* and *7* but it had been extensively modified.

To begin with, it had been fitted with a retro-rocket using hypergolic fuels to enable it to be placed in orbit round Mars, and two large propellant tanks had been added for the nitrogen tetroxide and monomethyl hydrazine. Hypergolic fuels are those which ignite on contact and therefore do not need

A close-up of the central body of Mariner 9.

an ignition system. The camera systems were also upgraded. The wide-angle camera was fitted with eight interchangeable filters, while there was a fixed yellow filter on the narrow-angle camera. There were five other experiments apart from the TV imaging system—an infra-red spectrometer, an infra-red radiometer, an ultra-violet spectrometer, S-band occultation and celestial mechanics. During the cruise to Mars the mass of the spacecraft was 975 kg and after the propellants had been spent in placing it into orbit round the planet this was reduced to 520 kg.

Mariner 9 travelled for 157 days in interplanetary space, needing only one mid-course correction on 5 June 1971, to direct it precisely on course for its aiming point near Mars. Then, on 14 November it became the first man-made artificial satellite of another planet. Its orbit round Mars was inclined at 64.28° to the Martian equator and had a 'periapsis' (low point) of 1,397 km, an apoapsis (high point) of 17,916 km, and a period of about 12½ hours.

But the elation felt by the team at Pasadena was turned to dismay when the first television pictures began to come back and it was realized that the entire planet was covered by a gargantuan dust storm. This had begun to be appreciated before *Mariner 9* arrived through studies by Earth-based astronomers, but the true extent of the obscuration was seen only when the

The Mariner 9 *spacecraft, the first to orbit Mars.*

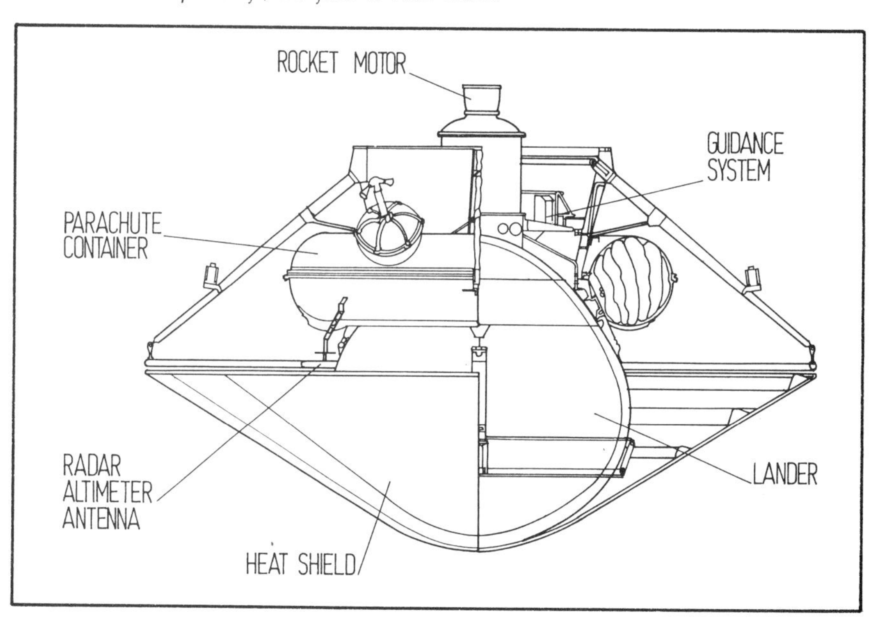

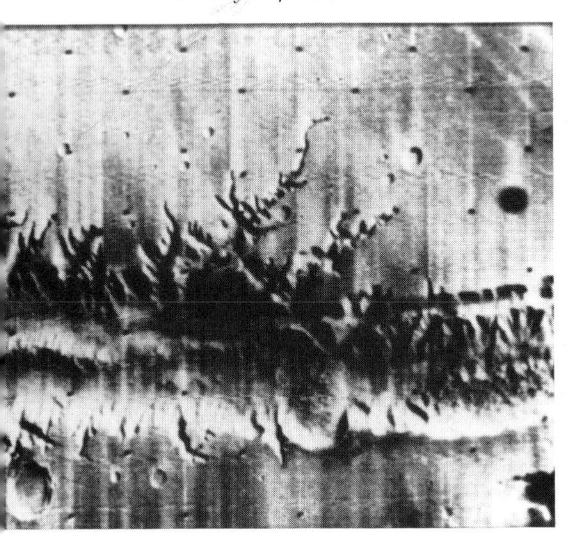

Left *Valles Marineris, the huge canyon in the northern hemisphere of Mars, viewed by* Mariner 9.

Right *Winter winds cause this pattern of clouds to the lee of a 90km crater pictured by* Mariner 9.

Below *A* Mariner 9 *picture of a Martian canyon near Tithonius Lacus, with a profile of the slopes.*

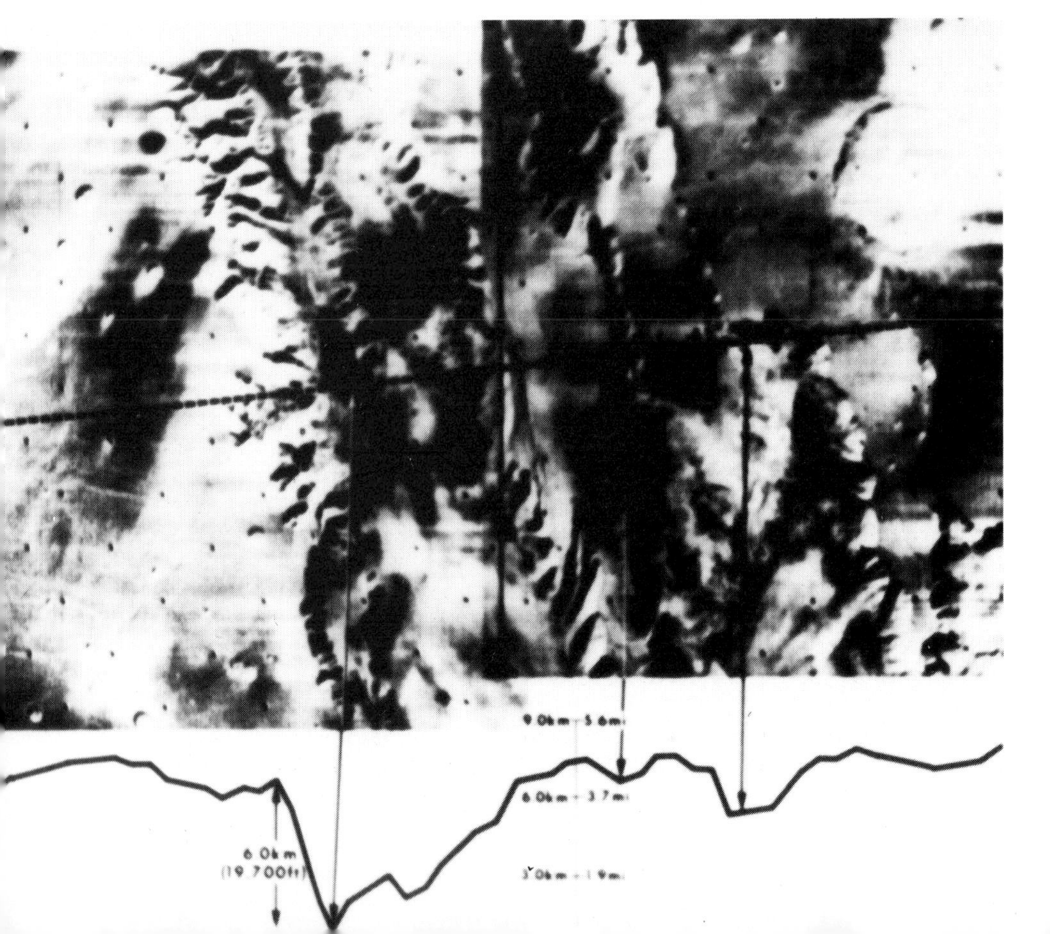

spacecraft's television system was turned on some time before orbital insertion. Until the middle of December all that could be seen through the dust cloud was faint markings and sometimes a diffuse feature followed by billowing dust clouds on its lee side. Pictures of the limb (edge) of the planet were analysed and showed that the dust reached an altitude of 70 km.

A 'T' pattern of dark spots did appear on the early TV images and the theory that they were high spots on the Martian surface showing through the dust was put forward. Eventually, the dust cleared enough to reveal that they were in fact the summits of four enormous volcanoes. Each of the volcanoes covered an area as big as the state of Arizona and the biggest of them stood 29 km above the plain. This was christened Olympus Mons and remains the outstanding feature on the surface of Mars.

Olympus Mons is an immense shield volcano with a crater at its summit 70 km in diameter. Its base is 500 km across and it is instructive to compare its dimensions with those of Earth's largest volcano, the island of Hawaii, which is 200 km across on the ocean floor and rises to a height of 9 km above the seabed. Olympus Mons, like the other shield volcanoes on Mars, is on a ridge which itself stands about 5 km above the mean level of the Martian surface. The other three volcanoes, Ascraeus Mons, Pavonis Mons and Arsia Mons, are only slightly less awesome than Olympus Mons and taken together show many of the features of terrestrial volcanoes. Erosion has been much less effective in the thin, dry atmosphere of Mars, however, and in some other respects the volcanic regions bear a resemblance to areas on the Moon.

Previous *Mariner* flights had led to the erroneous conclusion that Mars was a dead planet, devoid of geological, or more correctly areological, activity.

Mariner 9 overthrew this theory and it now seems that the planet's entire surface may have been affected and even formed by such activity. Certainly, the extensive plains, pock-marked as they are by impact craters of all sizes down to the limits of resolution, were formed by lava flows and ash falls.

More than 7,000 pictures were transmitted by *Mariner 9*, but probably none was more spectacular than those that showed the huge system of canyons in the equatorial region. This system stretches for 5,000 km and was named Valles Marineris in honour of its discoverer. But its origin remains a mystery. No convincing explanation of how the huge system of parallel canyons was formed has emerged. A minor side canyon in the system is roughly the same size as the Grand Canyon in Arizona. Great gulches up to 200 km wide stretch for thousands of kilometres and are up to 6 km deep. The major question at the time of the discovery was where the material has gone that was removed in their excavation.

Other prominent features revealed by the pictures included immensely long faults and fractures in the Martian crust, cliffs up to 4 km high and stretching for hundreds of kilometres, chaotic landscapes of jumbled rocks, and many thousands of craters, both volcanic and others formed by impacts. Some of the very large, multi-ringed craters appear to be almost as old as the planet and have been extensively eroded.

Unlike the Moon, Mars has at least a tenuous atmosphere and many of the *Mariner 9* pictures show the effects of wind erosion and deposition. The equatorial region, it transpired, is mainly under the influence of erosion, the wind lifting dust and sand and scouring away hills and cliffs and etching parallel grooves in flat areas. Nearer the poles, however, deposition of the wind-borne material dominates in the form of sand dunes and featureless spreads of dust.

Mariner 9 solved at least one Martian mystery—the origin of the so-called 'wave of darkening'. In the summer hemisphere of the planet astronomers have noted from Earth observations that there is a gradual darkening from the pole towards the Equator. A tentative explanation before the Space Age was that vegetation was responding to the release of water from the melting polar cap. But this was disproved by *Mariner 9*, which showed that the effect was entirely due to dust being removed by seasonal winds.

This was not the only changeable set of features seen on Mars, which appears to be by far the most volatile planet after the Earth. The polar caps of either carbon dioxide or water ice shrink and grow with the seasons and above all the clouds of Mars gave fascinating information to meteorologists. At the surface the Martian atmosphere has the temperature and pressure of our atmosphere 30 to 40 km above the Earth. But even at these extremes carbon dioxide and water can freeze into crystals, so clouds do appear. In the lee of large craters and other prominent features the phenomenon of lee waves was often seen. This is a type of cloud well known in mountainous regions of the Earth.

So the picture of Mars painted by *Mariner 9* is one of a strange world by

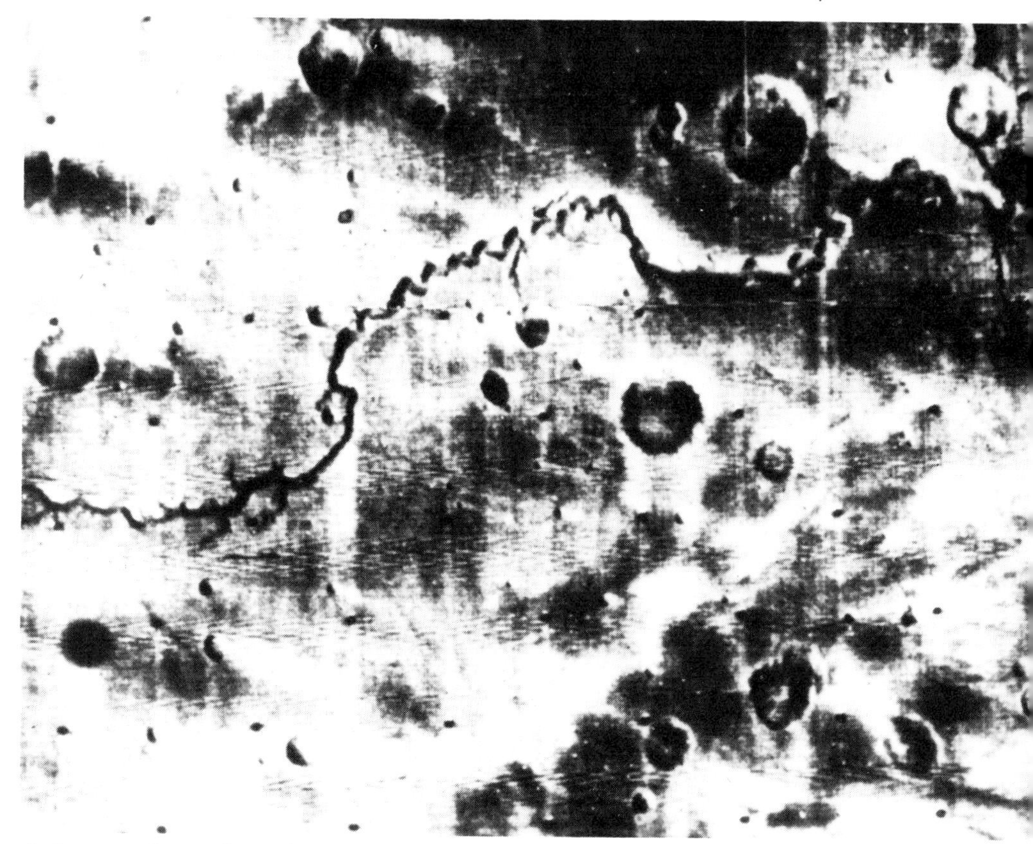

A sinuous valley on the surface of Mars — a water channel or a lava flow? Mariner 9 *did not provide the definitive answer.*

terrestrial standards. The first human explorer to go there will confront a world of high cliffs, immensely long canyons, volcanoes of unbelievable size, dried river beds and jumbled rocks, all in an eerily thin atmosphere where sound will scarcely carry. Reddish landscapes will shiver beneath pinkish skies decorated with occasional wispy clouds. It is not as strange as some of the fantasies about Mars created by science fiction writers and serious scientists alike, but it will be a truly alien world.

One final aspect of the *Mariner 9* mission was of great interest to astronomers. It was programmed cleverly to take close-up pictures of the two satellites of Mars, Phobos and Deimos. At least one well-known astronomer had postulated that these little moons, which have retrograde orbits round their parent, might be artificial. But the *Mariner 9* pictures showed that they are irregular bodies, shaped rather like potatoes—heavily cratered and

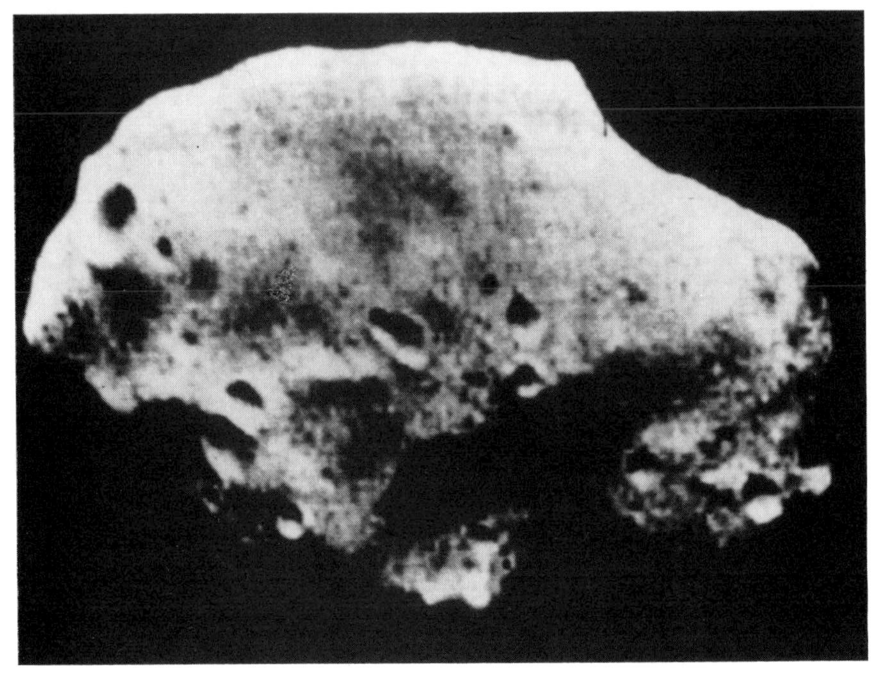

Phobos was photographed for the first time in close-up by Mariner 9.

natural. *Mariner 6* and *7*, back in 1969, had been the first to photograph the satellites. As they were on simple fly-by missions, however, they could not be targeted to take close-up, high resolution pictures of *Mariner 9* quality. *Mariner 9* was a highly successful exploratory mission which showed just how much more sophisticated American technology was than the comparable Soviet effort.

The Russian Mars craft which orbited the Red Planet at about the same time were four times as massive, yet as far as one can tell from published data, returned only a fraction of the useful scientific intelligence provided by *Mariner 9*. It was also a useful forerunner of the *Viking* missions which were despatched to Mars two launch windows later and which were as successful in their way as the *Voyager* flights to Jupiter, Saturn and Uranus.

Up to this time all the Mars probes had been strictly aimed at geophysical research into the anatomy of Mars and its atmosphere. With the *Viking* programme, the NASA team at JPL were trying for the first time to examine the biology of Mars, if there was any. For the first time, the primary aim of a space mission was to discover if life existed on another planet. Scientists are still arguing about the results, but it is certain that from the operational point of view the two *Viking* spacecraft were an outstanding success.

Chapter 7
The search for life

Viking 1 and *Viking 2* were highly sophisticated craft which were each equipped with an automatic laboratory capable of detecting chemical activity associated with life as we know it. But the *Viking* project was far more ambitious than this—it also embraced wide-ranging meteorological, geological, chemical and seismological studies, as well as a mapping programme that eventually photographed almost the entire surface of Mars.

Like all the other American interplanetary missions, Project *Viking* began at Cape Canaveral. The launch vehicles were the powerful Titan-Centaur combination and the *Viking* craft began their ten and eleven month voyages to Mars on 20 August 1975, and 9 September 1975, respectively. Only slight mid-course corrections were necessary, thanks to the accurate trajectories initiated by the launchers. On 19 June 1976, the retro-rocket of *Viking 1* was fired to put the whole spacecraft—orbiter as well as lander—into a highly elliptical orbit around the Red Planet. Periapsis was 1,500 km and apoapsis 50,600 km, while the inclination to the Martian equator was 37.8°. There was now no particular hurry about setting the lander down on the planet's surface and this took place on 20 July 1976. The landing site was in the plain known as Chryse Planitia, at a point 22.27° north of the equator and at a longitude of 47.97°. (Longitude is measured from an arbitrary meridian agreed on by international astronomers.)

Viking 1 was at a distance of 350 million kilometres from Earth when the lander separated from its 'mother' craft. Seven minutes after separation a compact computer in the lander issued a series of commands to small rocket thrusters on the heat shield. The motors fired for more than 22 minutes to reduce the lander's velocity and throw it out of its orbit around Mars.

Two radar altimeters fed altitude and velocity information to the computer which had to be virtually autonomous at this stage since commands from the control centre at the JPL would have taken nineteen minutes to reach the spacecraft. Eventually, the instruments in the lander began to sense the fringe of the Martian atmosphere and 3 minutes 59 seconds later, when the craft was 6 km above the surface, a big red and white parachute was deployed. This slowed down the lander from 830 kmh to 188 kmh and seven seconds later the heat shield was jettisoned. The last few metres of the

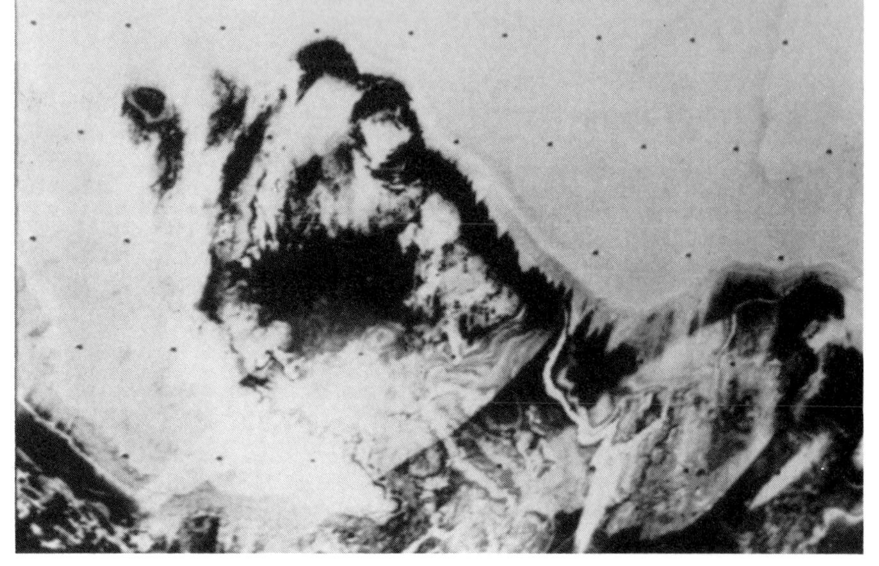

Great ice cliffs in the north polar region of Mars, photographed by a Viking *orbiter.*

descent was cushioned by small rocket motors and the *Viking 1* lander touched down in a dusty, rock-strewn plain. Minutes later the lander's television camera relayed back to Earth the very first picture ever taken on the surface of Mars—a shot of its own footpad resting on rocky Martian soil. The epic exploration of our most intriguing planetary neighbour had begun in earnest.

The excited scientists assembled at the JPL to watch the first pictures coming back from the Martian surface soon had much to discuss. From Chryse Planitia the *Viking 1* lander transmitted images of a fairly flat but undulating landscape of ochre-coloured material, strewn with boulders lying between small dunes. The dunes were, of course, of windborne material and other wind effects were also obvious. They included trails of fine grains lying between the boulders, carried there by prevailing winds in quite consistent directions. The entire landscape did not, however, owe its morphology to 'aeolian' erosion and deposition. There were visible stretches of a hardened crust of a vitrified nature, which demonstrated a process of evaporation of water which had left behind a layer of mineral salts. This was a landscape not unlike the stony deserts of North Africa, North America and Asia though devoid of the hardy desert plants to be found on Earth.

Meanwhile, *Viking 2*, on a slightly longer transfer orbit, was approaching Mars and in due course, on 7 August 1976, its rocket was fired to place it in a 1,502—35,728 km orbit around Mars: this time the inclination of the orbit was 55.6°. A site at 44°N had been chosen but *Viking 1* pictures showed that this site was situated in extremely severe terrain. In the next few weeks no less than 4.5 million square kilometres of the Martian surface was scrutinized before a new site was chosen. As a result, on 3 September 1976, the

Viking 2 lander was set down some 7,500 km from *Viking 2* in the area known as Utopia Planitia. The co-ordinates of the site were 47.67° north and 225.71° longitude.

The second lander found a stony plain stretching in all directions. The larger boulders and pebbles in the field of view may have originated as debris from a meteoric impact crater about 200 km away. Between the rounded boulders, which have a spongy, vesicular appearance typical of some terrestrial lavas, appeared a polygonal pattern of channels.

Viking was a large craft by previous American interplanetary standards. The orbiter weighed 2,325 kg before insertion into Mars orbit, and 950 kg after the fuel was exhausted. The lander was 1,090 kg on being ejected from the orbiter, and after its rocket had fired its weight on the ground was 600 kg—or rather, would have been under terrestrial gravity.

On board the *Viking 1* lander the computer had been pre-programmed to command the trench-digging arm which was part of the sampling equipment to deploy at a given azimuth, elevation and extension. When the first black and white and then colour pictures came back from the landing site they were examined closely and it was concluded that the chosen point was too dangerous because of numerous large rocks. So the computer was sent a signal to re-position the digging and the unique task of sampling Martian soil began soon afterwards.

There were three *Viking* experiments which were aimed at proving or disproving the existence of life or its remains on Mars. The first was the pair of cameras mounted on the lander to take black and white, colour and stereo pictures. They would have pictured any large life forms or plant growth, but they failed to do so. The GCMS (Gas Chromatograph/Mass Spectrometer) could have found organic molecules in the soil but to the surprise of the scientists at JPL it did not. Carbon, hydrogen, nitrogen and oxygen in the form of organic compounds are present in all living matter on Earth and simple compounds have been found by astronomers in many parts of the Universe. The GCMS searched for these compounds either as evidence of life, its precursors or its remains but found none.

The biology experiment, the third attempt to detect life, consisted of a small box containing three instruments which JPL described as the most sophisticated scientific hardware ever built. The objective of these instruments was to search for signs of metabolic processes such as those used by bacteria, green plants and animals. In one section, Martian soil was exposed to water vapour for six Martian days and then wetted with a complex solution of metabolites which would act as a nutrient to bacteria. The gas above the soil was then monitored by gas chromatography and oxygen was detected 2.8 hours after humidification. There was also an increase in carbon dioxide and nitrogen. In the pyrolytic release experiment, the soil was exposed to carbon monoxide and carbon dioxide in the presence of light. A small amount of gas was found to be converted into organic material, but heat treatment of a duplicate sample prevented such conversion.

Finally, in the labelled release experiment soil was moistened with a solution containing several organic compounds labelled with radiocarbon. Radioactive gas was released, but again heat treatment prevented this occurrence. The general conclusion was that the existence of life in the Martian soil was not proved, although not all scientists believe this to be demonstrated beyond doubt.

The orthodox view is that the release of oxygen from the soil indicated that it contained oxidants, which on Earth perform the role of breaking down organic matter and living tissue. The chemical reactions that took place in the Viking instruments were unexpected and unusual but they were not evidence of life forms. All the elements essential to life are present on Mars but liquid water is also needed and it can only exist on Mars now either as ice or as water vapour.

The conditions now known to exist on or just under the surface of Mars do not allow carbon-based organisms to exist and function. But the case for life in the distant past is still open, for the pictures from *Mariner 9* and the even more extensive coverage by the *Viking* orbiters revealed that liquid water did exist in large quantities on Mars in the past. The evidence is in the form of dried-up water courses, ranging in size from very large to very small, which showed that both rivers and spring-fed streams were common at one time.

Analysis of the soil by the two landers showed that silicon and iron are the most abundant elements. About 45 per cent of the soil consists of silicon dioxide and 19 per cent iron oxide, much of it in the form of maghemite, which undoubtedly gives the planet its characteristic red colour. Other elements present are magnesium, calcium, sulphur, aluminium, chlorine and titanium. There is 100 times as much sulphur as is found in terrestrial soil, but there is much less potassium. The high iron and low potassium have profound implications for the geology of Mars, which has probably not suffered as much differentiation of the elements by internal heating as the Earth has.

The general conclusion was that the soil might be a mixture of argillaceous minerals high in iron and iron hydroxides, together with other minerals high in iron and carbon. About one per cent of the soil by weight was water. Part of the water may be in hydrated minerals—on Earth such minerals produce clays rich in iron and the same thing could have happened on Mars when there was plenty of water on the planet.

With its record of dust storms and violent winds, the Martian surface had been expected to be a dangerous place for the *Viking* landers. But the scientists monitoring the landing sites were somewhat surprised by how little they changed. Only minor changes were recorded by the cameras. There were slight variations in the brightness and colour of a few places on the surface where dust was moved in thin layers and a couple of tiny landslips near the first lander.

The most significant change was deemed to be the appearance of a thin layer of what seemed to be water ice on the groundd near *Viking Lander 2*. This was a puzzle because in the Martian winter the atmosphere is so dry that it cannot hold enough water vapour to produce frost. The explanation may lie in water vapour adhering to dust grains in the warmer summer hemisphere and drifting north across the equator. When the temperature fell to the level at which carbon dioxide in the atmosphere froze, it condensed on to the water and dust particles and bore them down to the ground.

The landers made one momentous discovery about the atmosphere—it consists of 2.7 per cent of nitrogen, one of the essential elements of carbon-based life. No previous spacecraft had detected nitrogen in the predominantly carbon dioxide atmosphere. A seismometer on the second lander found no evidence of quakes on Mars, but meteorology instruments measured air temperatures, wind speeds and directions. They were able, in fact, to compile the first extra-terrestrial weather reports in the history of meteorology. *Viking Lander 2* functioned until 11 April 1980, but the first craft on the surface carried on working, sending back weather reports and pictures, until November 1982.

Viking *took this picture of the Tharsis Ridge on Mars, the planet's youngest volcanic region.*

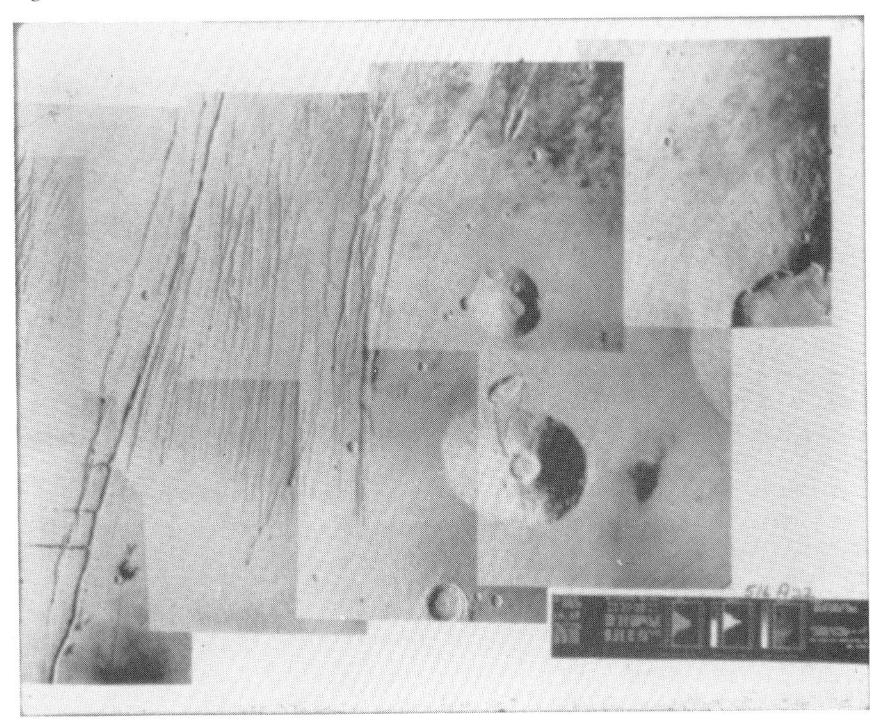

While the landers were attracting the majority of the attention, especially for their glamorous search for life, the orbiters were conducting their own programme of research. In addition to their cameras, with which new and startling pictures of the variegated landscapes of Mars were captured, the orbiters carried an atmospheric water detector which mapped the atmosphere for water vapour, an infra-red thermal mapper which measured temperatures, including seasonal changes and their own radios which were used to measure the density of the atmosphere by the distortion of the signals as they passed through it.

The thermal mapper provided new information about the polar caps. When *Mariner 4* returned a temperature of -62°C for the polar caps it was concluded that they contained frozen carbon dioxide, or 'dry ice'; the temperature was simply too low for water ice to be present. *Viking* confirmed that this was true of the southern pole, but in the north its measurements showed that the carbon dioxide ice is a winter phenomenon and that in the summer the polar cap does consist of water ice. The four telephoto lens cameras on board the two orbiters brought a new dimension of clarity and accuracy to the mapping of Mars. After their first task of finding suitable landing sites for the landers had been completed they went on to map virtually the entire surface of the planet and more than 51,000 images came flooding back to the JPL control room through the Deep Space Network over the years of operation.

The pictures gave a complete picture of the topology of Mars, showing that most of the southern hemisphere is higher than the average surface, while most of the northern is lower. There are exceptions: the large, deep basins in the south known as Argyre and Hellas lie far below the mean surface. The northern hemisphere has three raised areas, the Syrtis Major Planitia, the Elysium Mons volcanic province and most notably of all the Tharsis Ridge. This gigantic 'continent' measures 2,500 by 1,900 km and includes the mammoth Olympic Mons and three other volcanoes first observed by the cameras of *Mariner 9*. The ridge covers a quarter of the entire Martian surface and its bulge probably began to rise 3.3 to 4.1 billion years ago. The volcanoes formed much later and Olympus Mons probably began to erupt 2.5 to 3 billion years ago. When it did it must have been one of the most spectacular events in the history of the Solar System, for the lava from its crater cascaded down its slopes and over a great escarpment not much less than 6 km high to the plains below.

All the systems were still in operation when *Viking Orbiter 2* finally ran out of its attitude control gas on 25 July 1978. With no attitude control it was impossible to point the spacecraft at the targets, so the mission came to an end on that date. *Viking Orbiter 1* soldiered on beyond the end of the seventies and was finally declared dead only in the late summer of 1980. It was the end of an historic era in planetary exploration, but it overlapped with the *Voyager* era, which brings the story of planetary exploration up to date.

Right *Saturn, in a composite picture, with six of her moons.*

Below *False colour image of Saturn's rings.*

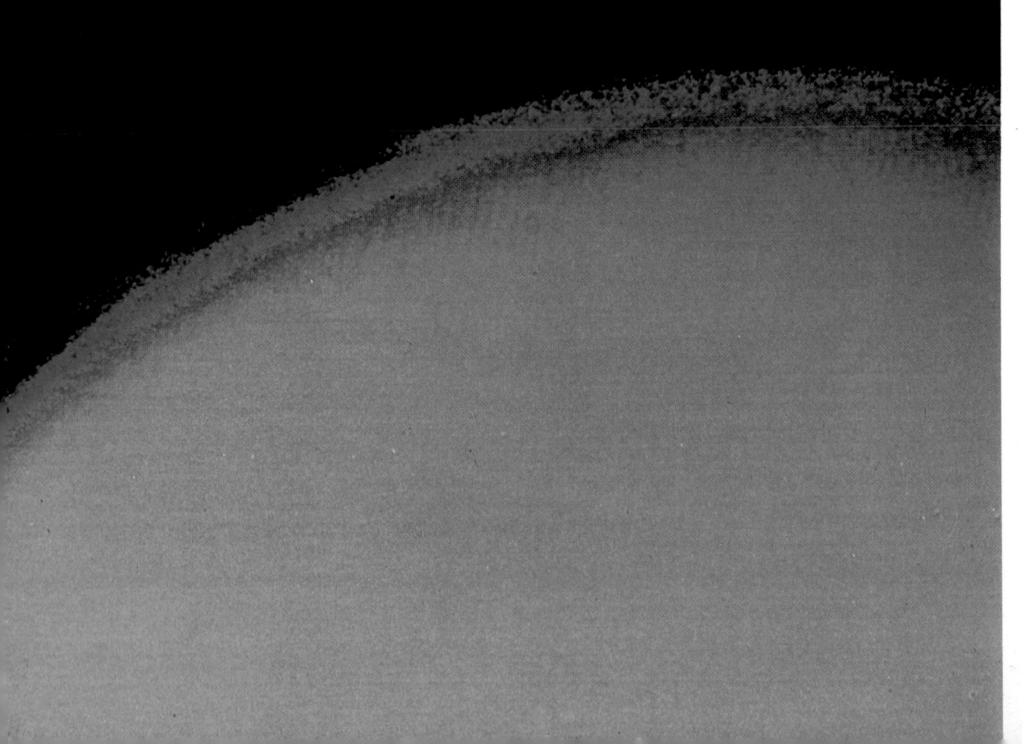

Left *The 'spokes' discovered by* Voyager *on the B ring.*

Below left *A thick haze above the atmosphere of Titan, one of* **Saturn's** *moons.*

Right *Tethys and its huge canyon system.*

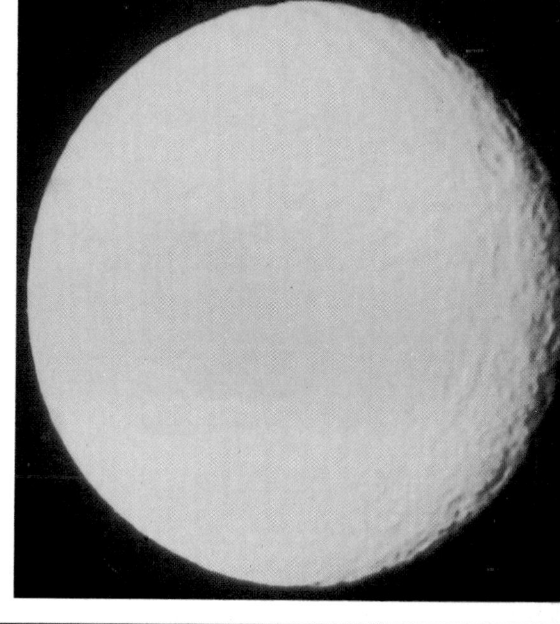

Below *The heavily cratered moon Mimas.*

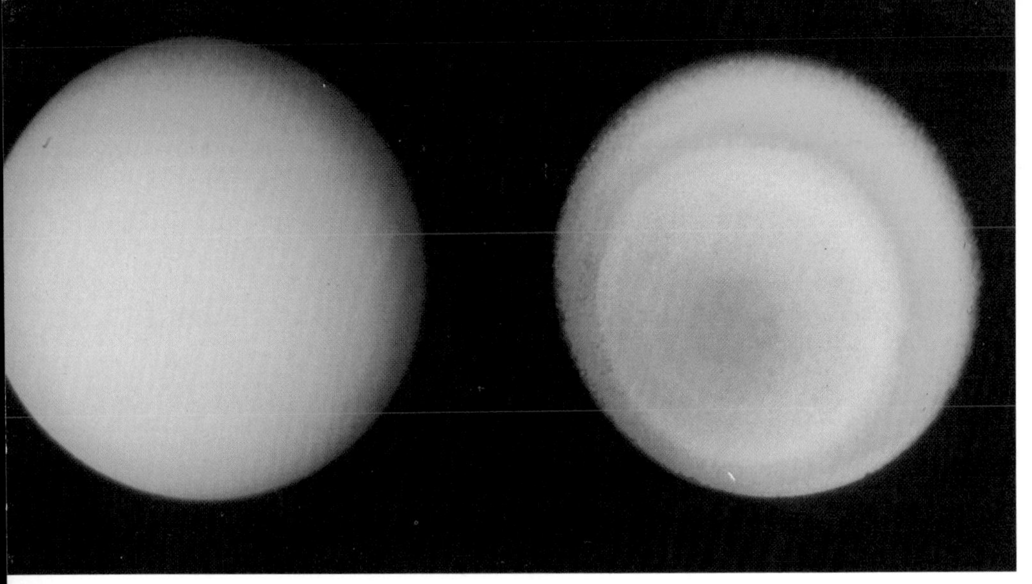

Above *Uranus in real and false colour.*

Below *Two tiny moons 'shepherd' a Uranian ring.*

Right *All nine known rings of Uranus.*

Below right *The battered face of Miranda.*

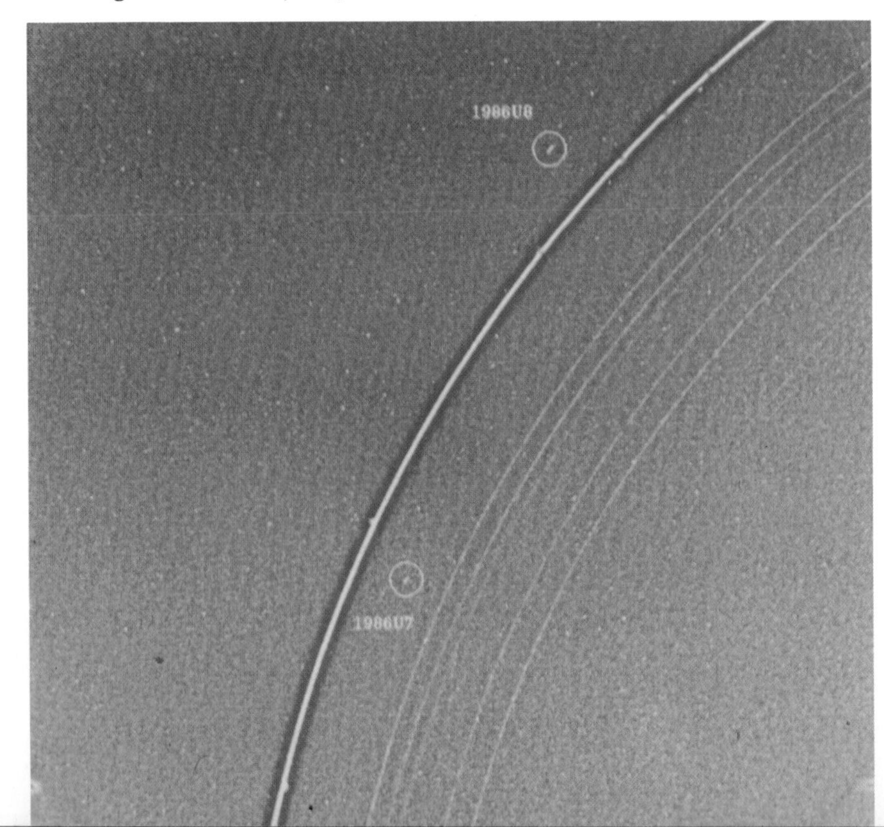

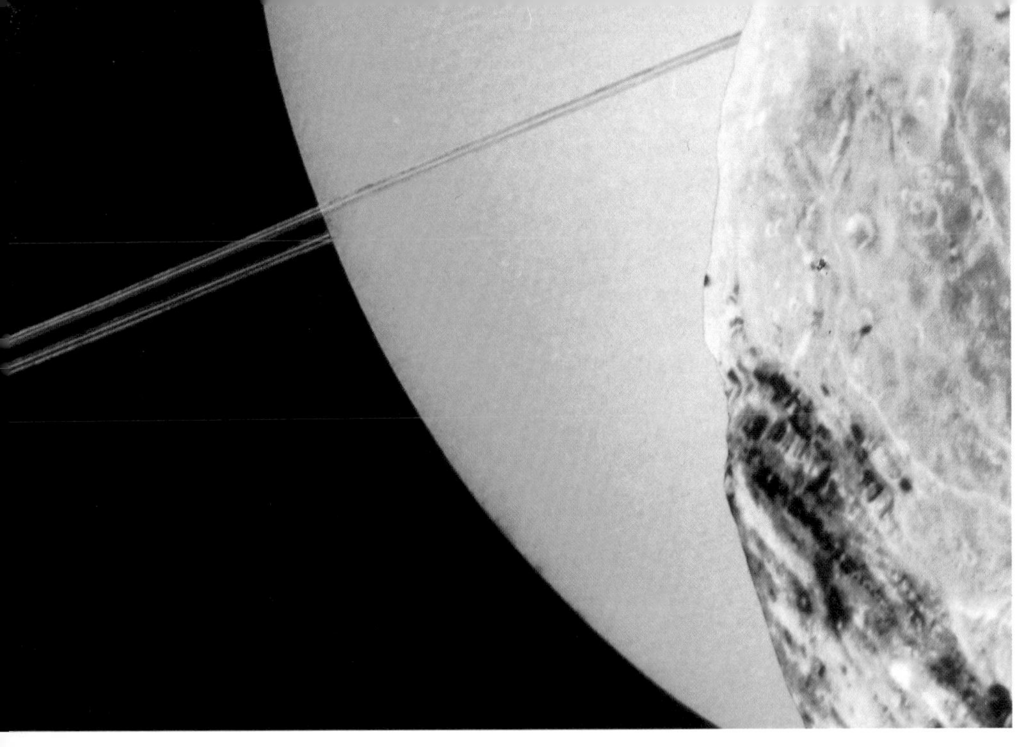

Above *Miranda in the foreground of Uranus.*

Below *The* Surveyor *spacecraft.*

Giotto *is launched from Kourou.*

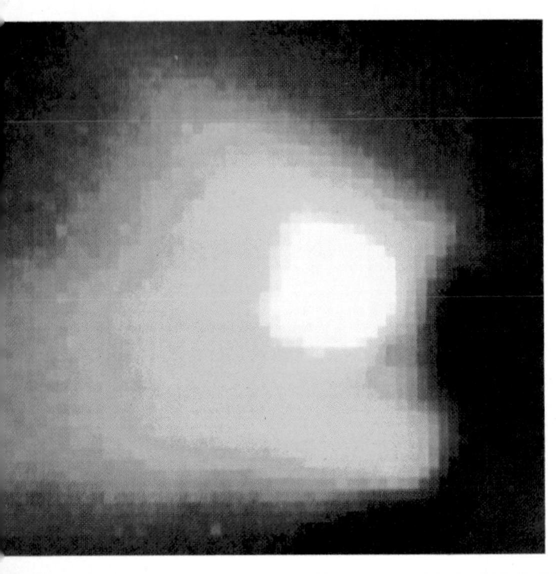

Left *A false colour close-up by* Giotto *of Halley's Comet.*

Below *Photograph of the Martian surface by* Mars 5.

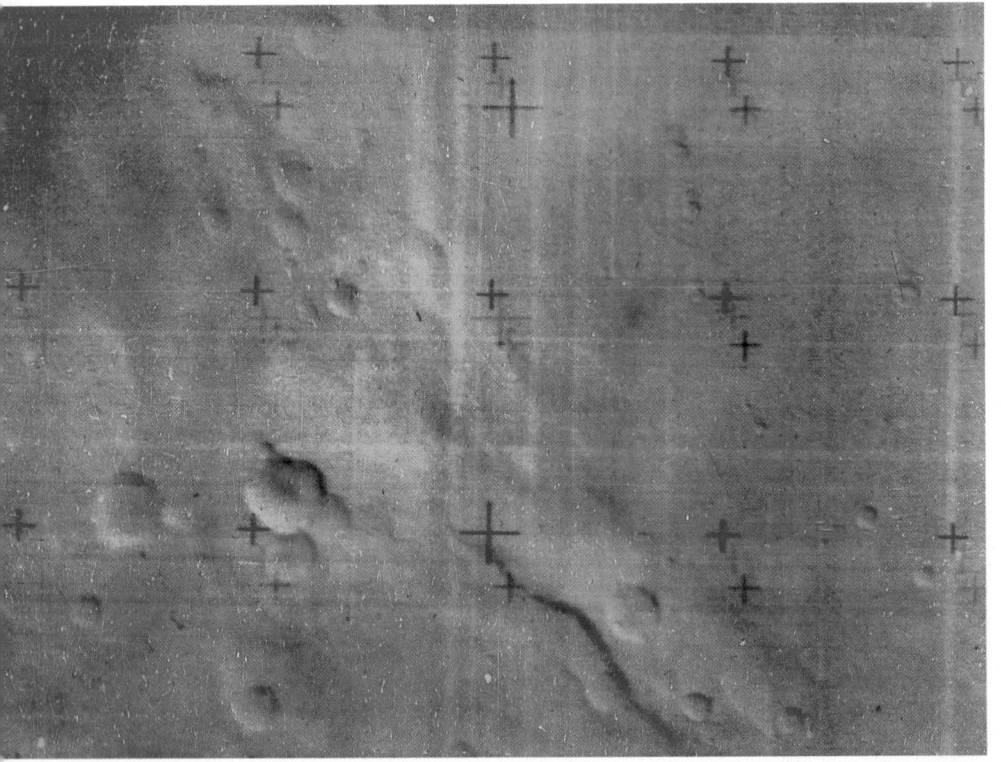

Chapter 8
Beyond Mars

Before going on to describe the remarkable *Voyager* missions, we must go back a few years to the time when America embarked on a bold new venture to explore the Gas Giants—the great array of planets from Jupiter outwards which make up the bulk of the mass of the Solar System. Perhaps the two little craft that are still speeding away from Earth were simple by later standards, but they produced incredible advances in scientific knowledge for minimum cost and prepared the way for *Voyager*.

The era of the exploration of the planets beyond Mars began on 3 March 1972, when the Atlas-Centaur rocket carrying *Pioneer 10* into space lifted off from a launch pad at Cape Canaveral. This tiny craft, with a mass of only 258 kg, started then an epic journey which is still going on. And it will go on, perhaps, for ever, or at least until the component parts crumble to dust under the hail of radiation which it will have to endure in interstellar space.

For *Pioneer 10*, guided by an ingenious team at the Ames Research Centre of NASA in California, and boosted by a third stage consisting of the solid rocket used on the *Surveyor* programme, became the first man-made object ever to attain a sufficient velocity to escape for ever from the gravitational attraction of the Sun. And to mark its unique role as man's first ever messenger to the stars it was equipped with a metal plate depicting in graphical form our place in the universe and the shape of the dominant species on our planet. This was a precaution just in case *Pioneer 10* was discovered in interstellar space by some intelligent race in the far distant future. There were critics who thought that perhaps we should keep our location secret, just in case any possible aliens turned out to be hostile and came visiting us with mayhem in mind. But less cautious counsel prevailed, especially as there is real doubt whether man will still exist on Earth in a million or more years from now.

Pioneer's real purpose was much closer at home, although still distant from its home planet by our puny standards. Its role—to become the first spacecraft to study the biggest planet in the Solar System, Jupiter, at close quarters. Jupiter is colossal by any standards. It is the biggest of our companions in the Solar System and was quite rightly identified in the Roman pantheon with the mightiest god of them all. To begin with, it is more mas-

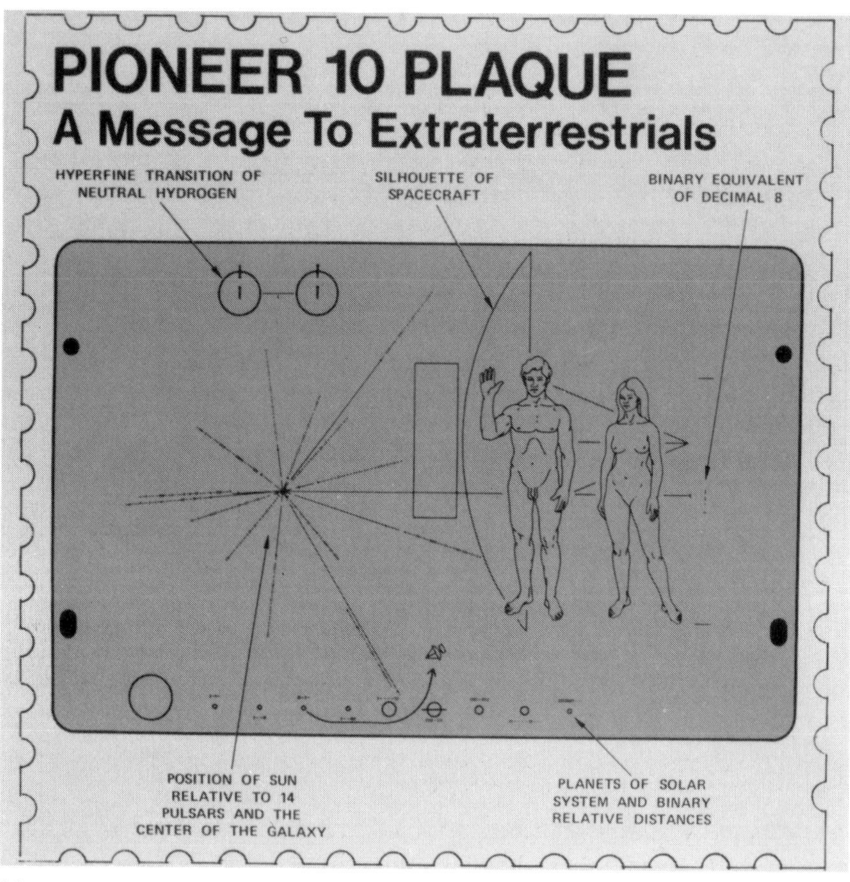

Message to extra-terrestrials. If any alien race finds Pioneer 10 *or* 11 *wandering through the Galaxy, this plaque will tell them where the spacecraft originated and what sort of creature made it.*

sive than all the rest of the planets put together and is equivalent in mass to 318 Earths. Although it is not as bright as Venus, which is so much closer, it outshines the other planets and when telescopes first began to be trained on it 370 years ago, astronomers found much to wonder at. Galileo himself is reputed to have been the first to see the four so-called 'Galilean' moons that orbit Jupiter—a discovery which destroyed for ever the myth that all heavenly bodies revolve round the Earth. Early telescopic images revealed that Jupiter has a unique pattern of bands around its globe, together with a Great Red Spot that has remained an enigma until the 20th century.

To study this mighty body, *Pioneer 10* was equipped with the best instruments that modern science could devise. As well as detectors designed to

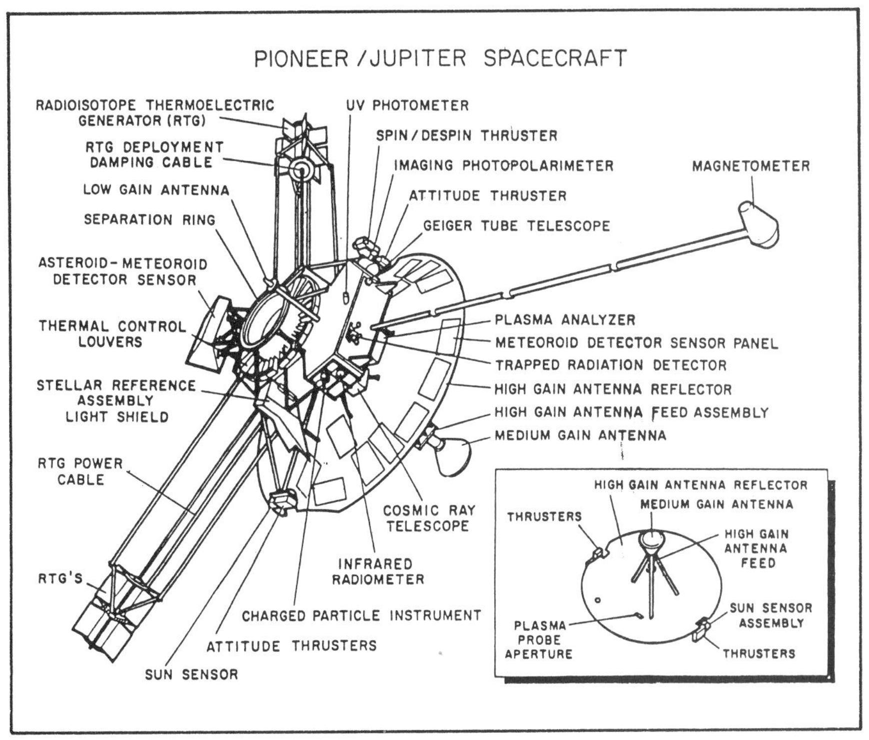

Above *The* Pioneer *spacecraft was a miracle of miniaturisation, packing an amazing amount of scientific equipment into its small bulk.*

Below *The RTGs which supplied power out beyond the orbit of Mars are seen on the short booms in this picture of* Pioneer 10.

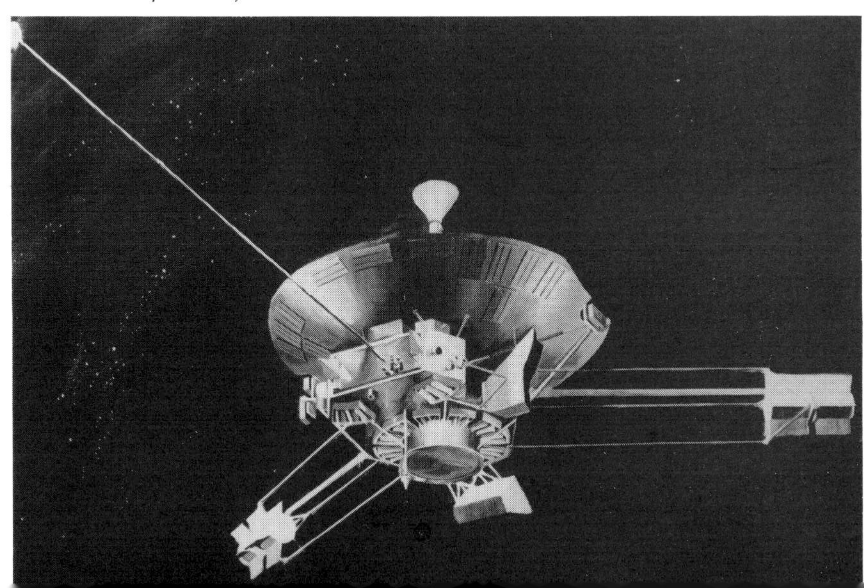

study cosmic radiation, ultraviolet, infra-red, plasma, charged particles and meteoroids, the spacecraft was equipped not with a camera but with what was called an 'imaging photopolarimeter'. With the weight limitation imposed by the launcher, a television camera would have been too heavy, so the lighter photopolarimeter was substituted for it. This was an instrument devised at the University of Arizona which was capable of making two-dimensional maps of the planet and its satellites showing the intensity and polarization of red and blue light by the use of filters. This technique did not provide the spectacular images of the planet supplied later by the *Voyager* spacecraft but hundreds of images were returned to Earth, showing the detail of Jupiter's zones with greater resolution than the best telescopic pictures.

Solar panels had been used for all planetary spacecraft to date, but these are not effective beyond the orbit of Mars because of the increasing distance from the Sun. Consequently the *Pioneers* became the first spacecraft to be powered entirely by nuclear energy. They were equipped with four Radioisotope Thermoelectric Generators (RTGs) in which the heat of the natural radioactive decay of plutonium-238 is used to generate electricity. At the beginning of the mission the power output was 155 watts, declining to 140 in the vicinity of Jupiter. Of this tiny output only 8 watts was directed into the radio transmissions of *Pioneer,* and this was reduced to an infinitesimal fraction of a watt by the time it was picked up by the Deep Space Network.

Pioneer 10, a spin-stabilized spacecraft (as distinct from the three-axis stabilized craft like *Mariner* and *Voyager*), given a velocity of well over 50,000 kmh by the three-stage rocket, made a fairly fast passage across the Solar System and arrived in the vicinity of Jupiter in December 1973. As viewed from Earth, it passed behind Jupiter on the fourth of that month, and at closest approach was only 132,252 km above the planet's colourful cloud tops. The images from the photopolarimeter showed features not seen clearly before, if at all, such as four bright spots about 4,000 km in diameter in the south temperate latitudes.

Further north there were several bright swirls of high contrast, and there was a striking wave pattern at some of the zone boundaries. Perhaps the most impressive feat of the *Pioneer 10* mission was the demonstration that despite an environment so hostile that a man would die in it instantly, man-made instruments could survive and operate successfully. When it reached the neighbourhood of Jupiter, *Pioneer 10'*s instruments reported back that it was bathed in huge clouds of highly energetic electrons, protons and helium nuclei. The radiation dose to which it was being subjected amounted to 200,000 rads from electrons and 50,000 rads from protons. For a man, a whole body dose of only 500 rads is lethal.

A few small problems with instruments occurred while this deadly radiation persisted, the most serious of which was a series of false signals to the imaging photopolarimeter which led to a failure to carry out a planned imaging of the Jovian satellite Io. But all the instruments subsequently

Above Pioneer *flies past Jupiter, its antenna pointing back at the Earth.*

Below Pioneer 10 *reaches Jupiter, an artist's impression which can give only a faint idea of the grandeur of the vista seen by* Pioneer's *imaging device.*

returned to full working order and the spacecraft was still broadcasting at full power more than four years after it passed Jupiter. Only then did the increasing distance make *Pioneer 10* inaudible from Earth, and by that time the little craft had crossed the orbit of Uranus.

Pioneer 10 discovered that Jupiter, which is an extremely energetic giant, has a magnetic field 250,000 times as great as Earth's. This undoubtedly accounts for the lethal radiation belts that surround the planet. It also found that the centre of the field, or 'magnetic dipole', is not at the centre of the planet but 18,000 km removed from it. The axis of the dipole is at an angle of 15° to the axis of rotation. Another major discovery, although one that had been suspected from Earth observations, is that Jupiter radiates away into space 2½ times as much energy as it receives from the Sun. It has even been said about Jupiter that it is a star waiting to happen. If it was very much bigger the thermonuclear reactions which power the stars would begin and the *Pioneer 10* results certainly show that it is in a highly energetic state.

One of the techniques used in the *Pioneer* mission, as well as on later flights

Pioneer 11 *at Saturn. No TV camera could be carried because of the weight restrictions but* Pioneer 11 *nevertheless returned marvellous images and contributed significantly to knowledge of the planet.*

The awesome sight expected to confront Pioneer 11 *if it approached too close to Saturn's rings — a maelstrom of rocky and icy fragments.*

to the planets, was to obtain details about the atmospheres of the main bodies and their satellites by analysis of radio tracking data. This 'radio science', by detecting changes in radio waves passing close to Io, enabled the *Pioneer* team to conclude that this interesting satellite of Jupiter has an extremely tenuous atmosphere, and also an ionosphere. It is a remarkable feature of the radio science that this study of Io was possible for it depended on extremely accurate guidance. *Pioneer 10,* which went behind Jupiter as seen from Earth, also went behind, or 'was occulted by', the tiny satellite—a great tribute to the mission planning and to the level of guidance accuracy achieved.

Io is interesting particularly because it is implicated in the intensely energetic radio emissions which make Jupiter the strongest radio source in our sky after the Sun. Its passage through the planet's radiation belts is evidently the cause of these natural radio broadcasts which have been studied from Earth for some forty years. Even before it reached Jupiter the craft had carried out crucial reconnaissance of an area of space which some experts thought might be lethal to man's probes—the asteroid belt.

Strewn between the orbits of Mars and Jupiter are some relatively large minor planets and untold quantities of large and small rocks, dust and microscopic particles. It was feared that a spacecraft passing through them

A demonstration of the effect of computer enhancement. The upper picture was received from Pioneer 11*'s close encounter with Saturn; the lower picture has been computer enhanced to bring out more detail.*

would be damaged or destroyed. *Pioneer 10*, by surviving unscathed, demonstrated that this was not so, a fact that three succeeding spacecraft have confirmed. Instruments on *Pioneer 10* measured the incidence of micrometeoroids and showed that their concentration was not high enough to endanger even a high-speed interplanetary traveller. It had been calculated that a particle weighing a thousandth of a gramme would penetrate a centimetre of aluminium, orbiting the sun as it does at a speed of 61,200 kmh in the middle of the asteroid belt. Most of the particles detected by the two *Pioneers* were smaller than this and fewer particles than predicted were encountered. The conclusion was that dangerous concentrations of high velocity dust which could endanger spacecraft simply did not exist in the asteroid belt, so the JPL scientists who were planning the *Voyager* missions were able to take heart from the experience of their colleagues at the Ames Research Center.

The slingshot effect of Jupiter's gravitational field placed *Pioneer 10* into a trajectory which will take it into interstellar space, arriving near the star Aldebaran in about 1.7 million years. But the Ames team planned an even more exciting future for its successor, *Pioneer 11*. It was decided to take *Pioneer 11* even closer to Jupiter, in fact to 43,000 km above the cloud tops, which involved a passage right through the radiation belts. This was not as dangerous as it might sound, for the chosen trajectory approached Jupiter from below the south pole, so *Pioneer 11* would hurtle almost straight up through the radiation belts. This reduced the time the craft would spend within the lethal area and in the event it received a smaller dose of radiation than *Pioneer 10*. This particular trajectory was chosen because of the exciting additional mission of *Pioneer 11*—to explore the planet Saturn. Even with its powerful three-stage launcher, the craft could not possibly have reached the ringed planet without the extra boost given to it by the gravitational field of Jupiter. But the influence of Jupiter would accelerate *Pioneer 11* to 55 times the speed of a rifle bullet—173,000 kmh—and that would be enough to propel it halfway across the Solar System to Saturn after a journey of 3,800 million km. It had the added bonus of throwing the little craft into an arcing trajectory no less than 164 million km above the ecliptic plane, into a region of space which had not been explored previously. Interplanetary craft usually travel in or close to the plane in which the Earth revolves around the Sun, since it takes excessive energy to fly out of that revolution.

Pioneer 11 was duly launched on 6 April 1973, and the slight increase in its velocity which made the close approach to Jupiter possible was commanded a year later, on 19 April 1974. This was just after it had emerged from the asteroid belt, unscathed like its predecessor. At its encounter, it sent back excellent real-time pictures of Jupiter which were far better than any ever obtained by Earth-based telescopes and actually won for Ames Research Center an 'Emmy' award from the National Academy of Television Arts and Sciences. Several instruments showed anomalous readings because of the radiation they were subjected to, but most recovered completely. The two

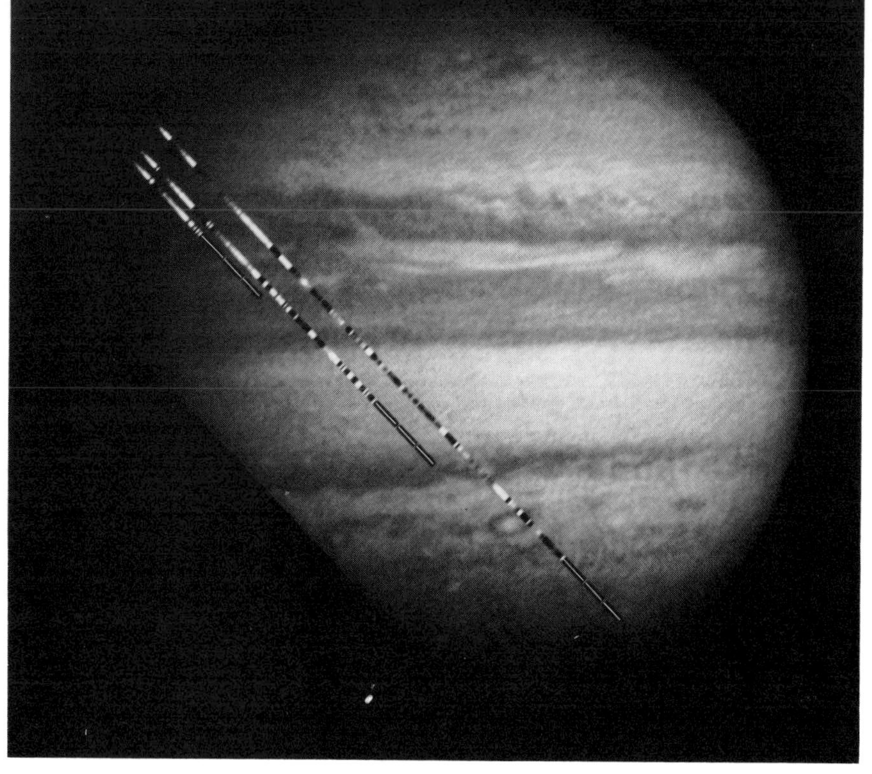

Pioneer 10 *took this picture of Jupiter. The diagonal lines result from errors in transmission.*

Pioneers between them returned about 1,000 pictures of the Jovian system, including the first close-ups ever taken of the four Galilean satellites, Io, Europe, Ganymede and Callisto.

Although the satellite pictures suffer by comparison with those later returned by *Voyager 1* and 2, at the time they represented a great technological feat. *Pioneer 11* continued on its journey and on 1 September 1979, duly passed within an amazing 20,200 km from the cloud tops of Saturn. In passing through the system of the planet, the craft came within 1 million km of no fewer than nine of Saturn's satellites, Phoebe, Iapetus, Hyperion, Dione, Mimas, Tethis, Enceladus, Rhea and Titan.

As well as returning the best pictures of Saturn to that date, *Pioneer 11* was able to study the structure of the famous rings, the dynamics and climate of the atmosphere, and the interaction of the planet's magnetic field with the solar wind. Inevitably, as the *Voyager* results flowed in through the Deep Space Network during the next two years, the success of the *Pioneer* programme began to be forgotten. But the two little craft were true pioneers who paved the way for their much more advanced—and expensive—successors and by the data they gathered made their successors' passage easier.

Chapter 9

Voyagers to the stars

Of all the spacecraft that have been sent speeding towards the planets in the last three decades, none deserves the plaudits of the world more than the two *Voyager* probes. All the superlatives apply to them—they have been successful beyond the wildest dreams of their creators. They have opened up a succession of new worlds both to scientific study and to the curiosity of the public and they have done more than any other group of craft to clarify some of the questions about the Solar System that remained to be answered at the dawn of what has come to be called the Space Age. Of course, they have also returned results which have posed many new questions, but that is inevitable in any line of scientific inquiry.

Project *Voyager* has been all the more remarkable because it was a substitute programme, a second best which came about because a more grandiose mission to the outer planets was abandoned before it began for reasons of economy. This was what was christened 'the Grand Tour'—a concept which was never finally authorized by the US government. The Grand Tour was conceived by American scientists as a once in two or three lifetimes

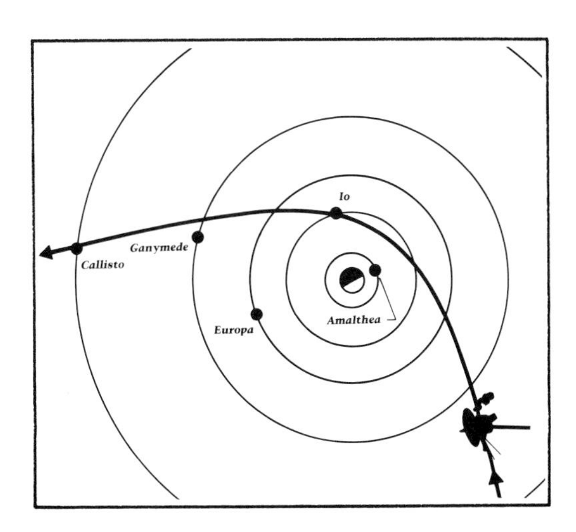

Voyager 1, *in its historic encounter with Jupiter, was also able to take close-up pictures of Io, Ganymede and Callisto.*

opportunity to explore the outer planets at a minimum cost. The possibility arose because once every 172 years the outer planets, from Jupiter to Pluto, line up on one side of the Sun in such a way that a single spacecraft projected on the correct path would fly close to each one in turn. The planets are not in a straight line, but in a curve which would place each one of them in a position to be visited in turn by a man-made probe from Earth.

One of these opportunities occurred in 1979 and all through the sixties there was excited talk in the American scientific community about the possibility of a Grand Tour mission in that year. Scientists at the Jet Propulsion Laboratory even went so far as to design a self-repairing computer which would be given virtually autonomous control of such a mission, since the lapse between sending a signal from Earth to a spacecraft two or three billion kilometres away, then waiting for a reply to confirm that it had been received, can run into hours, even at the speed of light. The idea that the Grand Tour spacecraft should have a 'brain' of its own, capable of putting right its own faults, was a clever one, but the ultimate solution developed for the *Voyager* mission was just as clever in its own way.

Eventually, however, the cost of the Grand Tour mission militated against its acceptance by the administration, and it seemed that the opportunity to explore the outer planets might be missed. But the JPL team concluded that much of the Grand Tour could still be carried out by a spacecraft based on the successful concepts of the *Mariner* missions to Mars and Venus. And that was how *Voyager* was born—a much more ambitious piece of engineering but firmly rooted in the experience that was giving new views of Mars and Venus through the sixties and the seventies.

When it came to designing the *Voyager* craft JPL had a new set of con-

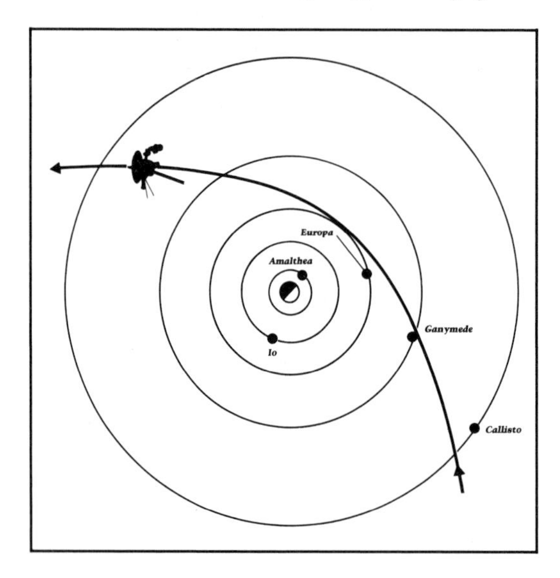

In its encounter with Jupiter, Voyager 2 *passed close to Callisto, Ganymede and Europa.*

straints to bear in mind. They had to operate over a wide range of environmental conditions and for a longer period than any previous spacecraft. The distances from Earth over which they would operate made it imperative that communications capability was enhanced, that the hardware was more reliable, and that navigation and temperature control were more sophisticated than ever before. JPL met these challenges with an expertise that has become legendary in the space community.

One of the wonders of the *Voyager* missions is that so much was packed into a total weight of only 825 kg, of which the instruments made up only 117 kg. It was a feat of miniaturization and design which has few parallels in the brief history of space exploration, and it is a great tribute to the team that designed and built them that the two *Voyager* probes are still operating in space almost a decade after they were launched. If all goes well *Voyager 2* in particular will more than vindicate its creators' faith by exploring yet another virtually unknown planet, Neptune, twelve years after its launch, in 1989.

At the centre of the ungainly-looking craft is a ten-sided aluminium framework with ten compartments for instruments and equipment. It weighs 24.5 kg and is 47 cm high and 178 cm across. In the centre of the decagon is a spherical tank containing 104 kg of the propellant, hydrazine, which is used in twelve small thrusters which control the attitude of the craft and four which are employed for course corrections. Mounted on the decagon on struts is the high-gain antenna, the all-important dish through which all communications to and from Earth pass. The 3.66 m diameter dish is an aluminium honeycomb structure surfaced on both sides with laminated graphite-epoxy. Literally billions of data points have passed back and forth through these two antennas since the *Voyager* probes were launched, although compared with the planets they have visited and the space they have crossed they are totally insignificant specks in the Universe.

There are three booms attached to the central decagon, giving *Voyager* its characteristic 'three-legged spider' look. The science boom, comparatively short at 2.3 m long, and made of graphite-epoxy tubing, carries on its end a scan platform which can be steered in two axes, so making it possible to point cameras and other instruments at a passing planet while the high-gain antenna is still pointing faithfully back at Earth. Two TV cameras, one wide-angle and the other narrow-angle, two spectrometers for studying infra-red and ultra-violet emissions, and a photopolarimeter are on this platform because they are most sensitive to radiation and it is as far away as possible from the radioactive source which powers the spacecraft. A source of electric power other than the solar panels used on *Mariner* was essential because the energy received from the Sun is affected by the inverse square law—go twice as far from the Sun and the power received is reduced to a quarter; three times as far away and it is reduced to a ninth, and so on. The chosen solution was the same as that adopted in the *Pioneer* project. Solar panels simply could not generate enough electricity to power the *Voyagers* out among the

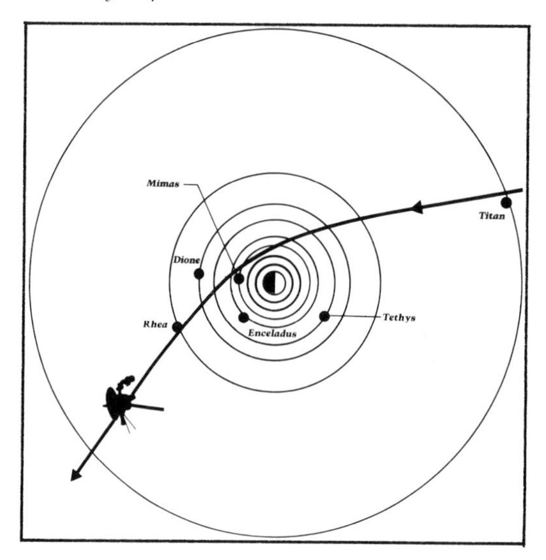

Voyager 1 *passed just outside the rings of Saturn and made many discoveries about the Saturnian system.*

outer planets, hundreds or even thousands of millions of kilometres from the Sun: each spacecraft was therefore fitted with three radio-isotope thermoelectric generators. At the beginning of the missions 160 watts was available from each of the three RTGs.

The output declines gradually over the length of the mission. To keep the plutonium as far as possible from delicate instruments, the RTG units are on a short boom opposite the scan platform. Since even a small spacecraft has its own magnetic field, the *Voyager* experiments to detect low magnetic fields needed to be remote from the hardware, so two magnetometers are mounted on a 13 m epoxy glass boom. Two high field magnetometers are mounted on the spacecraft itself. Other instruments, concerned with cosmic rays, charged particles and plasma-science, are scattered along the science boom. The experimental array is completed by two 10 m whip antennas which are part of the planetary radio-astronomy experiment.

The delicacy with which a tiny spacecraft has to be controlled at a distance of billions of kilometres from Earth can be contemplated, if it can scarcely be pictured. At the heart of this feat of scientific wizardry is the computer-command subsystem (CCS), which gives the spacecraft a semi-automatic capability, although not as sophisticated as that planned for the Grand Tour. The time lapse in communications, a consequence of the finite speed of radio waves means that there is no possibility of an Earth controller commanding the spacecraft to perform its functions in real time. There has to be an element of stored commands and the CCS was designed to carry out this method of control. In the CCS are two independent memories, each capable of storing 4,096 'data words'. Half of each of the memories stores reusable fixed routines that do not change during the mission, and the other half is programmable by updates from the ground.

Most commands to other sub-systems in the spacecraft emanate from the CCS memories, which at any given time are loaded with the sequence appropriate to the particular phase of the mission. There is an accurate clock on board and the CCS counts the time until a preprogrammed interval has elapsed and then branches into sub-routines that result in commands to other subsystems. When a phase of the mission is approaching, such as a fly-by of a planet, the Deep Space Network sends an update to the CCS controlling all the functions which thrusters, the scan platform and the instruments have to perform. This update may cover a period of hours during a planetary encounter, but during the cruise phase, between planets, it may last for months. At the appropriate times the CCS issues commands to the attitude and articulation subsystem (AACS) for platform activity or spacecraft manoeuvres or to the flight data system (FDS) for changes in instrument configuration or the rate of telemetry return.

Because of the distances and complexity involved in spacecraft operations the JPL controllers have developed ingenious methods of compressing the commands which are sent to the *Voyagers* in updates. A single 1,290-word science sequence load, for instance, can easily generate 300,000 discrete commands. It was necessary to send eighteen-sequence memory loads to *Voyager 1* for its encounter with Jupiter, and to convert the desired science objectives and measurements into the computer code which could be loaded into the CCS took the designers of the mission almost two years. Accuracy of the complex codes being uplinked to the spacecraft is of the utmost importance. There have been occasions in the past when a single digit inaccurately loaded into a computer has led to mission failure, so it is a remarkable tribute to the JPL team that nothing like that has happened during almost a decade of the life of the most complex unmanned spacecraft so far flown. The art of space navigation has also been refined during the mission.

American spacecraft from *Mariner* onwards have used a system of navigation which relied on the Sun and Canopus, a bright star in the southern sky. The principle is that if one electro-optical sensor locks on to the Sun and another on to Canopus the spacecraft can be stabilized in all three axes by sending signals through the AACS to the small thrusters. Once the spacecraft is stabilized it is possible to use extremely sophisticated methods to determine its position and trajectory and to change the latter when necessary. The primary method of making this determination is called radiometric observation. This is achieved by tracking the spacecraft with the antennas of the Deep Space Network in California, Australia and Spain.

Three quantities are measured by DSN. The first is the difference in frequency between the signals sent to it and those received from it—the so-called Doppler shift. It is this change in frequency which causes the best-known example of Doppler shift, the change in pitch in a train whistle as it passes an observer. Doppler measurements provide the navigator with a precise measurement of the velocity of the spacecraft away from the tracking station. The time a signal takes to come back to the DSN also gives the range of the spacecraft from Earth.

The third measurement uses two Earth stations simultaneously to receive signals from the spacecraft. Comparison of the signals makes it possible to estimate the relative angle of the spacecraft to the line between the two stations. Calibration of the baseline between the two stations is achieved by similarly correlating the signals from a quasar—a quasi-stellar radio source well outside our own galaxy. *Voyager* spacecraft can be navigated to an accuracy slightly better than one part in a million using radiometric data alone. So at a distance of 1.6 million km the navigation error would be only 1.6 km, which is comparable to hitting a target the size of a tennis ball at a distance of 40 km. But, to take the latest planetary encounter as an example, Uranus was almost 3 billion km away from Earth in January 1986, and the navigation error would have been more than 2,500 km using radiometric data alone. This is not accurate enough, so the radiometric data are supplemented by optical navigation methods, using the Voyager's own cameras. Pictures taken by the high resolution TV camera showed Uranus's satellites against a background of stars, making it possible to reduce the error to a mere 88 km—equivalent to hitting an aspirin tablet at a range of 40 km.

Radiometric data come into their own again when the analysis of the trajectory is made after the planetary encounter. The planet's gravitational field bends the trajectory, so inducing large velocity changes which can be measured precisely as Doppler shifts, and as a result the trajectory through the Uranian system will be known with an error of only 1.6 km.

Although most of the attention is focused on the spacecraft during the spectacular parts of the mission, the ground-based sector of the whole system is by far the most important. Without the sophisticated operational and analytical methods and techniques developed by the scientists and engineers at the laboratory in Pasadena the *Voyager* probes would be nothing more than pieces of man-made junk wandering around the Universe. To be present at the Mission Control and Computing Centre at JPL during a planetary encounter is like being present as a bit player in a giant Hollywood spectacular. There is all the excitement of waiting for the news to come through of some epoch-making event, with the whirring of the computers and the thrill of watching for the first pictures to appear on the numerous TV screens. This scenario has been acted out on many occasions, with both the *Voyager* craft encountering Jupiter and Saturn and *Voyager 2* going on to Uranus. But it also happened much earlier, with *Mariner* and *Viking* spacecraft sending back previously unseen views of Mars and Venus. A day at the JPL watching new images return from deep space is one of the new experiences thrown up by the Space Age, comparable for the enthusiasts with waiting for the first pictures to come back from *Apollo 11* or the first live landing of the Space Shuttle.

Although it is only part of the whole mission, the picture-taking capacity of *Voyager* is the part that has most captured the public imagination. The TV cameras have worked amazingly well over the years and the communications subsystem has matched them by sending back many thousands of pictures of the highest quality. The system sends back pictures which consist of 800 lines each of 800 pixels, or picture elements, making a total of 640,000 pixels per picture, at a rate varying from well over 100 kbs (kilobits per second) at Jupiter to about 30 kbs at Uranus. The brightness level of each pixel is measured on a scale from 0 to 255 and is transmitted by the spacecraft as an eight bit number in the binary code. (In this code 00000000 is zero, and 11111111 is 255.) The raw picture data are computer enhanced to produce better contrast and more details of the distant targets.

As an example of the difficulties which the mission planners encounter in making sure that sharp pictures are taken, when *Voyager 2* encountered Uranus, its speed relative to the planet and its satellites was of the order of 60,000 kmh. Obviously the cameras cannot be simply pointed at a satellite or planet for the duration of an exposure, however short. At this velocity the spacecraft is travelling more than 16 km in a second and there has to be some compensation for the motion, since exposure times varied between about a second and half a second. The solution, target motion compensation, was actually devised at JPL after the launch of the two *Voyager* probes. The scan platform moves in small steps rather than in a smooth motion, so it could not be used to cancel out all the drift, and the final solution consisted of changing the software of the AACS so that the spacecraft remained locked onto a target as long as the exposure lasted. This, together with the later computer enhancement of the images, accounted for the crystal clear pictures of celestial bodies as far as 3 billion km from Earth.

At the distance of Uranus, the data compression techniques included an ingenious method of reducing the number of bits that had to be returned for each picture. The software was altered so that the CCS sent back not the brightness level of every pixel, but the differences between neighbouring pixels. This led to a 60 per cent reduction in the bits needed per image. Since the TV cameras are of the black and white variety it may be wondered how such startlingly successful colour pictures have been returned. The answer is that red, blue and green filters are used and it is then possible back on Earth to reconstruct colour images using information only from a black and white camera.

The epoch-making journeys of the two *Voyager* probes began, oddly enough, with the launch of *Voyager 2* from Cape Canaveral on 20 August 1977. This arrangement came about because *Voyager 1*, for reasons connected with trajectory choice, was to make a quicker dash to Jupiter and would arrive first. The launcher chosen for the project was the Titan, an obsolescent ICBM more powerful than the Atlas, with the trusty Centaur as the second stage. *Voyager 1* duly followed its sister-ship into space on 5 September 1977. The initial velocity was so high that each of the *Voyager* craft crossed the orbit of the Moon within ten hours.

Although the missions have been such a triumph, success was not completely unalloyed. There have been two major failures in the spacecraft, the second of which, the seizing up of the scan platform of *Voyager 2* , will be described later. The first fault, which gave JPL cause for much concern for many months, was the failure of the primary radio receiver on *Voyager 2* before the encounter with Jupiter.

What made the failure worse was that the back-up radio, which has had to be used ever since, lost its ability to tune automatically to the frequency of the transmissions from Earth. The frequency changes because of the velocity of the spacecraft relative to the Earth and also the temperature of the receiver. New techniques allowing engineers to predict the relative velocities and to transmit at the precise frequency which *Voyager 2* can 'hear' have overcome the problem.

Voyager 1, however, has experienced few problems and it was in full working order as it approached Jupiter during the first major post-launch event—the rendezvous with the biggest planet in the Solar System. It was on 5 March 1979 that *Voyager 1* made its closest approach to the planet, passing a mere 348,890 km from the top of Jupiter's cloudy atmosphere. But since the instruments and cameras are able to explore a planet at a great range, the encounter phase of the mission actually began much earlier. Measurements of the Jovian system began two months earlier and lasted until 13 April, a total of 98 days.

With the RTGs providing 445 watts of power, the *Voyager* was able to send back one picture every 48 seconds, and 3,600 bits of science and engineering data every second. As a result, it was able to return much new information about Jupiter's satellites, atmosphere and magnetosphere. From the public's point of view, perhaps the most interesting discoveries were that Jupiter, like Saturn, has a ring system, and that Io, the closest of the four Galilean satellites, has at least eight active volcanoes. The ring is a 30 km thick band of material around Jupiter, with an outer edge 128,000 km from the centre of the planet.

It was only when telescopes were first turned towards the skies in the seventeenth century that men realized that the bright planet Saturn was surrounded by an almost equally bright system of rings. For well over 350 years it was taken for granted that it was the only planet in the Solar System with such rings, and a planet with a ring system is often used by artists and cartoonists to illustrate the concept of a strange and alien world. But a flight over the Indian Ocean in an airliner specially adapted by NASA as an airborne observatory changed all that and brought the whole subject of rings and their origins to the attention of planetary scientists.

After the discovery from the Kuiper Airborne Observatory that Uranus had rings, the investigation of the phenomenon both there and around Jupiter became a topical subject for those planning the *Voyager* missions. The discovery was made in 1977 by Dr James Elliot. He and his co-workers wanted to study an occultation of a star behind Uranus in order to glean facts about the distant planet and its atmosphere. Occultations, in which a star

Dione, one of the moons of Saturn, showed numerous impact craters in this view from Voyager 1.

passes behind a planet or satellite, give scientists a unique opportunity to view the effect of an atmosphere on light travelling through it. Uranus duly occulted the star, but the watching scientists noticed that its light flickered five times before the occultation and five times after it. They had discovered that at least five rings surround Uranus, and the following year they found four more. There were some theoretical problems to be solved before the phenomenon was understood.

It was obvious that some gravitational effect was keeping the particles that made up the rings out of the gaps which lie between them, and it was important to find out why. The discovery was made six months before the launch of the first of the *Voyager* spacecraft and they were quickly re-programmed to search for a Jovian ring. In due course both spacecraft were able to confirm that such a ring did indeed exist at a distance from the planet equal to 1.8 times its radius. At this distance there is a known gap in the radiation belts that surround Jupiter and it had been reasoned by scientists at the Goddard Space Flight Center in Maryland that either an unknown satellite or a ring was absorbing the radiation there and so causing the gap. *Voyager 1* caught a glimpse of the edge of the ring, but *Voyager 2* was instructed to get a better view and revealed that the ring was 800 km wide and was flanked by dimmer rings. Scattered light was analysed to show that the particles making up the ring are extremely small—measured in millionths of a metre.

Jupiter's ring is unlike those which circle round Saturn, both in the size

Events in the *Voyager* programme

20 August 1977	Launch of *Voyager 2*
5 September 1977	Launch of *Voyager 1*
5 March 1979	*Voyager 1* encounters Jupiter
10 July 1979	*Voyager 2* encounters Jupiter
13 November 1980	*Voyager 1* encounters Saturn
26 August 1981	*Voyager 2* encounters Saturn
24 January 1986	*Voyager 2* encounters Uranus

of the particles, which are much larger in Saturn's case, and in their distribution. The Jovian ring does not cut off sharply above and below the equatorial line; instead, *Voyager* found particles both above and below the line, which may be due to the strong magnetic field surrounding the planet. The magnetic field lines go up and down through the rings because the magnetic axis is inclined at 15° to the axis of rotation. This may knock particles out of the rings. When the *Voyager* spacecraft went on to Saturn they found, like *Pioneer 11*, that the famous Saturnian rings were more complex than they seemed from Earth. In fact, they revealed that there are literally thousands of rings in the Saturnian system, and they are controlled by extremely complicated gravitational forces.

Pioneer was directed to fly outside the three known rings, known as A, B, and C, because of the suspicion that there was a previously unknown ring between them and the planet, presenting an incalculable risk of collision at 13 km a second. As it happened, the instruments on *Pioneer* detected a ring not inside, but outside, the three. It was called the F ring and turned out to be only 500 km wide, which is why it had not been detected before.

When the *Voyager 1* cameras were turned on it, the F ring was shown to be a much more interesting object than merely a thin band of particles orbiting Saturn. Those scientists who study celestial mechanics (ie, the way in which objects in space orbit round one another) expected these interactions to be fairly simple, but the F ring was far from simple. It had unexplained clumps and kinks and was braided into three separate strands. Further complication was to follow, for when *Voyager 2* arrived several months later the F ring had lost its kinky, braided appearance but was separated into five strands. But one of the ringlets within the A ring *did* present a kinky appearance.

Pictures of the C ring, which is closest to Saturn and a beautiful mixture of gold, orange, blue and white bands, showed a surprising spoke structure. These spokes, radiating out from the edge of the ring nearest the planet, move rapidly across the ring, sometimes as much as 20,000 km in twelve minutes. They fade as quickly as they appear and the scientists believe that they are imprinted on the rings in some way by Saturn's magnetic field. Particles charged by sunlight may be elevated above the plane of the rings by magnetism, so changing the characteristics of the rings' reflectance.

One of the disappointments of the *Voyager 1* encounter with Jupiter and Saturn was that the imaging photopolarimeter failed before it could be used. But on the *Voyager 2* mission the instrument worked perfectly despite some radiation damage and an ingenious use of it permitted a fine resolution of the Saturnian rings which exceeded that of the television cameras. As *Voyager 2* flew over the rings, the photopolarimeter monitored the light of a faint star, Delta Scorpii, viewed through the rings.

Since each ring blocked the starlight, and each gap allowed it through, the instrument was able to send back information about the complete ring formation across a distance of 70,000 km. It recorded that there are literally many thousands of ringlets and the F ring alone, for instance, is divided into ten components, of which the thickest is split into two minor bands only 200 m across. The ring edges are often extremely sharp, the A ring for instance fading away into nothing in less than a kilometre.

The theory which attempted to explain the sharp divisions and the many ringlets involved the 'shepherd and sheep' idea which postulated that the rings are kept in order by tiny moonlets. Their gravitational attraction on the trillions of particles in the rings, it was reasoned, keeps them in place. Only in the case of the F ring was it possible to prove that this was correct—the two minor satellites that act as shepherds for this ring were eventually detected. The other moonlets are probably too small to see on the available pictures.

In the case of *Voyager 2* the rings of Saturn were particularly important. Having duly made its close encounter with Saturn on 26 August 1981, and the navigation having been as precise as usual, *Voyager* passed within the planned 100,000 km or so of the ringed planet. The results—not only the dramatic pictures but the reams of scientific data as well—far exceeded the expectations of the 120 waiting mission scientists at Pasadena. After all, *Voyager* had already been in space for just over four years and had suffered a number of serious component failures. It would have been understandable if these problems had degraded the scientific value of the encounter, but this was not to be the case.

Yet there was to be one more crisis which, it seemed, might jeopardize not only the rest of the Saturn encounter but also the planned close approach to Uranus in 1986 as well. As part of the flight plan, *Voyager 2* had to disappear behind Saturn and while it was out of radio contact with Earth it also had to plunge through the plane of the rings. Since the rings are composed of trillions of pieces of rock and huge quantities of smaller particles, there was an obvious hazard to the spacecraft in this passage.

In the control rooms at the JPL champagne corks popped when the Deep Space Network picked up the infinitesimally faint radio signals emitted by *Voyager* on its re-emergence from behind the planet—still in one piece, it seemed. But the euphoria was quickly replaced by gloom when it became apparent that the craft's scan platform had stuck and could not be moved by command from Earth. This threatened to reduce the mission's usefulness during the rest of the flight through Saturn's system of rings and satellites and to make the Uranus visit impossible.

As explained earlier, the scan platform was designed to enable the scientific instruments and cameras of *Voyager* to point at targets of interest while the craft itself was stabilized in three axes, with the high gain antenna pointing towards Earth. Such an arrangement was necessary both because of the need to communicate at all times with the Earth and also because of the high speed of *Voyager* relative to the planets: at the Saturn encounter, for instance, its speed was 13 kms.

The platform is driven by a high-torque electric motor and it was believed at first that during the Saturn fly-by some contamination stuck in the motor's gears. But over the next few days some incredibly ingenious detective work established that this was not the case—another malfunction had caused the scan platform to stick. Fortunately, much of the imaging and science had been achieved by the time the platform seized up so very little was lost. Now ground tests on a prototype of the actuator that drives the platform, a spare that was not used in the spacecraft, and on numerous gear assemblies that were built to flight specification were undertaken at the JPL.

The platform had four different rates at which it could be slewed and the ground tests established that all the gear assemblies would seize up when they were subjected to the highest rate. This was apparently due to lubrication failure. The tests also showed that a 'healing' process took place in due course. New communications programmes were quickly devised to try to loosen the platform by running the motor in small steps. The sheer difficulty of communicating with a tiny craft a billion kilometres from Earth, let alone carefully and gingerly testing one of its components to try to correct a fault baffles the imagination. Although the radio signals giving instructions to the systems may be transmitted with the power of several thousand watts, by the time they reach the spacecraft their strength is reduced to only an infinitesimal fraction of a watt.

These weakest of weak signals, coded so that the *Voyager* computer can understand them, then have to be translated into actions—in this case switching the platform motor on and off. It is a great tribute to the skill and dedication of the JPL team that they were able to correct the scan platform fault. In fact, by 4 September, the scan platform was usable and *Voyager 2* was able to direct its instruments and cameras at Phoebe, the outermost satellite of Saturn, which revolves around the planet in a retrograde orbit, possibly because it was an independent body that was captured by Saturn's gravitational field.

The impact of the scan platform failure on the Saturn encounter was minimal. Edward C. Stone, the chief *Voyager* scientist, commented: 'We've accomplished most, if not all, of our major objectives.' The most spectacular of the acquisitions from the encounter were, as usual, the pictures of the planet, its rings and its satellites. On this occasion there were 17,000 of them, but the scientific data from the instrumental observations were, if anything, even more important to the assembled scientists.

Between them the two *Voyager* probes imaged all of the important satellites of Jupiter and Saturn as well as discovering some new minor ones. The satellites of the Gas Giants are by no meaans negligible either in their size or in their scientific importance in the understanding of the origin of the Solar System. Several are bigger than our own Moon and two, Titan and Ganymede, are bigger than Mercury and approach Mars in diameter, if not in mass. Another, Callisto, has a diameter only a few kilometres smaller than Mercury.

The discoveries made in the first high resolution images of these little worlds can only be summarized here. The most surprising, undoubtedly, was the existence of at least eight active volcanoes on *Io*, the closest of the Galilean satellites of Jupiter. Io has a diseased-looking surface, marked with volcanic calderas and lava flows in vivid reds and yellows. It is evidently a young surface which is constantly renewed by the volcanic eruptions which are stirred up by the close proximity of Jupiter. Io is the only body in the Solar System apart from the Earth where current vulcanism is proved, although it is now suspected on Venus (see Chapter 4).

Europa is another puzzling world. It has an outer icy coating, perhaps 100 km thick, round a rocky core. The surface is extremely smooth, with variations of relief only hundreds of metres high but the whole surface is covered with a maze of linear markings. The lines could result from the extrusion of fresh ice from below the surface into cracks caused by tidal forces resulting from the influence of Jupiter on the satellite.

By contrast, the biggest moon, *Ganymede*, shows several different types of terrain, evidence of a number of periods of geological activity. There are dark areas, heavily cratered during the formation of the Solar System 4.6 billion years ago, and lighter areas resulting from later geological activity which display many parallel lines of mountains and valleys. There are also newer craters which are surrounded by blankets of light-coloured ejecta. Ganymede is a strange world, made up of half ice and half rock.

Although much the same size, and also half ice and half rock, *Callisto*, the outermost of the Galilean satellites, is quite different. The surface is pock-marked with thousands of craters uniformly distributed and Callisto evidently underwent no further geological change after the initial bombardment of the various bodies which were orbiting round the Sun more than four billion years ago. The biggest impact has left a series of concentric rings on the surface. An inner satellite of Jupiter, *Amalthea*, proved to be an irregular-shaped object, too small to be shaped as a sphere by its own gravitational forces.

When the *Voyager* craft reached the Saturn system, *Titan*, which is almost as big as Ganymede, was imaged but no details of the surface were detected because the satellite has a mainly nitrogen atmosphere and sunlight falling on the top of it forms an impenetrable haze. But a dark ring at about 70° north was seen and also a lightening of the southern hemisphere; these may both

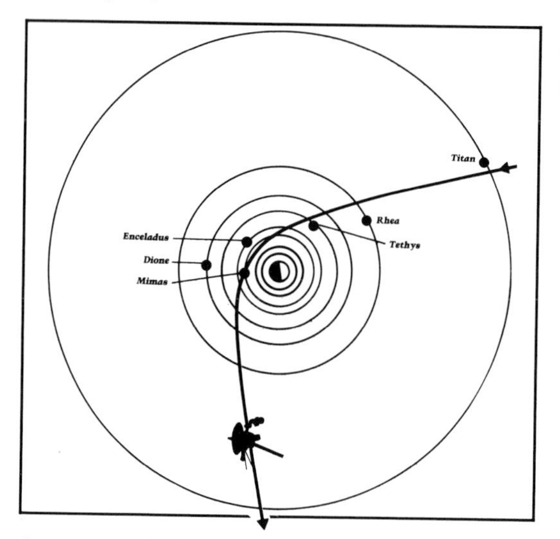

When Voyager 2 *reached the vicinity of Saturn it imaged six of its icy satellites.*

be seasonal effects. Titan may have lakes of frozen methane and clouds of liquid and frozen methane and may produce organic compounds which are the precursors of life, although it is thought to be too cold there for life to have evolved.

Water ice is a major constituent of the satellites of Saturn. One of them, *Mimas,* is heavily cratered and has one crater which is so large that it must have come perilously close to splitting the 392 km moon into fragments. *Iapetus,* 1,460 km in diameter, is half dark and half light—the dark material may be methane welling up from the interior. *Hyperion* is one of the most battered objects yet imaged in the Solar System and may be the remains of a large satellite rent asunder by some ancient collision.

One of the most distinctive surfaces was revealed on *Enceladus,* where five distinct terrain types have been identified. Craters are not as numerous as on some of the other icy satellites, and there are also smooth areas which have formed in the last billion years. There have been several eras of melting of the surface, and linear markings in the southern hemisphere are faults resulting from movements of the crust. A globe-encircling chasm distinguishes *Tethys,* which is also heavily cratered. The chasm is 2,500 km long and is probably associated with a 400 km crater on the other side of the 1,060 km globe.

Planetologists have acquired enough data from these amazing spacecraft to keep them happy for years, debating the lessons to be learned from data and images from more than a score of different worlds of varying sizes. They are not the sort of worlds where any sensible mission planner of the next century would aim to set up manned colonies but it is feasible that one day instruments will be landed on such moons as Titan to carry on the research that *Voyager 1* and *Voyager 2* have so ably begun.

Chapter 10
The blue planet

Voyager 2 had already been in space for almost 2,000 days when it made its closest approach to the most enigmatic target on its programme so far—the planet Uranus. The planet was an enigma, not because the mission scientists anticipated that something absolutely astounding would be found there, such as life or a new form of matter, but simply because the distance from the Earth that makes observation so difficult rendered Uranus an object of almost complete mystery. Like Jupiter and Saturn, Uranus is classified as a 'gas giant'. Like them, it has a core of silicate rocks and other heavy materials, surrounded by an atmosphere of hydrogen and helium, together with smaller quantities of other gases, in particular methane. But it is different in size, being only 14.5 times as massive as the Earth, whereas Jupiter is 318 times as large and Saturn 95 times. Facts such as these can be calculated from the orbits which the planets describe around the Sun and their perturbing effect on the orbits of other planets. In the case of Uranus, the period of its orbit is 84 years, but it is so far away, more than 3 billion kilometres, that it is impossible to see without a telescope, unlike Jupiter and Saturn, which were known to the ancients. It was not until 13 March 1787, that Uranus was discovered. Sir William Herschel, the British astronomer, was observing stars when he stumbled upon the 'new' planet.

In its remote orbit, Uranus is much less affected by the Sun's rays than is the Earth. The amount of energy from the Sun falling on a square metre of Uranus's surface is only one 400th as great as that falling on the same area of the Earth. In at least one respect Uranus is unique among gas giants. The inclination of its axis of rotation—running from north pole to south pole—is virtually horizontal. Inclination is an important factor in determining weather on a planet; if the Earth's axis was at right angles to the plane of its orbit round the Sun, for instance, we would have no seasons, but in fact the axis is inclined at 22½ degrees, which gives us our summers, autumns, winters and springs.

Uranus points its south pole directly at the Sun at this particular epoch. When it is on the other side of its orbit in forty years or so it will point its northern pole sunwards. One of the surprises of the *Voyager 2* fly-by was that this fact does not seem to have very profound effects, as astronomers had expected it to. Exactly why the rotation of Uranus is so abnormal is not understood, but it is believed that the planet may have been struck by a very large body early in its life, which knocked it over on its side.

Other facts known about Uranus in the dark age before *Voyager 2* paid its visit included the existence of five satellites, none larger than 1,600 km in diameter, a temperature of 60°K at the top of the clouds in the atmosphere and, most recently of all, a set of rings that match those round Saturn for interest if not for beauty (see Chapter 8). So it was not far from being an unknown planet when the veteran spacecraft approached it on 24 January 1986. As pictures arrived on Earth in the days preceding the closest approach its bluish hue, believed to be due to methane in the atmosphere, was seen.

No astronomical observations from Earth had detected the sort of banding that makes Jupiter and Saturn such ornamental objects in close-up and even in Earth-based pictures, but the imaging team back at JPL were delighted to see that Uranus too exhibits the same kind of banding, although much fainter than on the more conspicuous gas giants. The imaging was intensely difficult. Because the amount of sunlight falling on Uranus is so small and the distance so great, all the technical tricks perfected by JPL over two decades of planetary exploration had to be employed to collect the data and picture information returning to the world-wide Deep Space Network.

Voyager 2 had to be stabilized particularly firmly by newly-developed engineering routines and the computers had to be re-programmed to cram more data into each second. The result was a series of pictures of the planet showing the banding which were enhanced by computer techniques at JPL, demonstrating that the bands were essentially the same as those on the other gas giants. The banding is believed to be due to differences of temperature in the atmosphere. Where the temperature is higher gas wells upward, leading to high clouds of lighter colour, or haze. Neighbouring bands of darker colour are formed when colder gas sinks and the banding is caused by smearing round the planet by its high speed of rotation. The only other features seen were a faint brownish smudge on the south pole, probably a photo-chemical smog, and four clouds. The clouds were important because, as on Jupiter and Saturn, it was possible to measure the speed of jet streams in the upper atmosphere. On Uranus these reached 100 metres a second, compared with the 40 metres a second of terrestrial jet streams that help airliners cross the Atlantic more quickly. The Uranian clouds distinctly resembled thunderheads and it is possible that they were situated over gigantic storm centres. The direction of the winds detected on Uranus was interesting as well. It had been predicted that the winds would be blowing in a direction opposite to the direction of planetary rotation. This prediction was made because, as a result of the 90° inclination of the planet's axis, it was expected that the south pole would be 5° warmer than the equator. On a planet like Uranus temperature differences take centuries to level out, and it was expected that the north pole too would be warmer, even though it is now pointing away from the Sun. But infra-red measurements made by

The course taken by Voyager 2 *through the sideways-on 'target' of the Uranian system.*

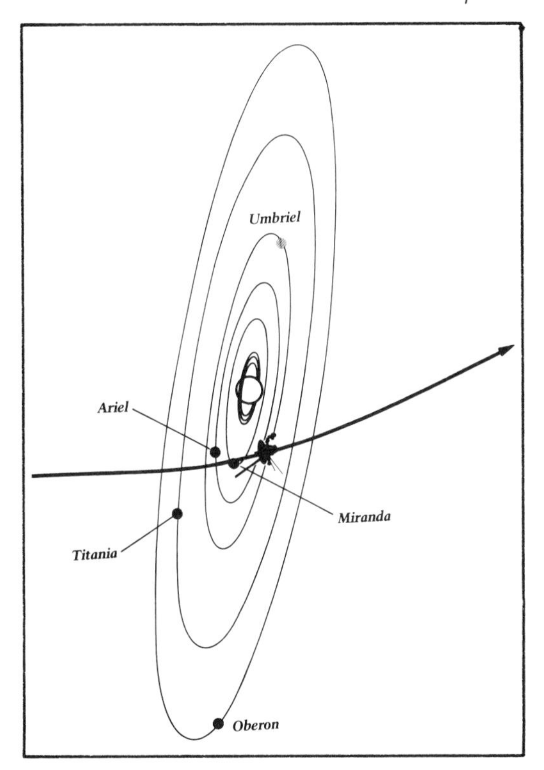

Voyager's instruments showed that there was no such variation in temperature and the winds were blowing *with* the direction of rotation. It seems that some process in the atmosphere is redistributing the heat received from the Sun uniformly over the whole planet.

When making the strategy for the Uranus fly-by the JPL planners were constrained in particular by the need to direct the spacecraft extremely precisely to ensure an accurate fly-by of Neptune in 1989. Uranus was literally like an archery target—lying on its side as it does with its rings surrounding it—and the trajectory could go through only one point on the target if it was not to miss its rendezvous with Neptune.

As it happened, this meant that the *Voyager's* cameras could be trained on only one of the planet's five known satellites for high resolution, close-up pictures. The fact that this satellite was the one known as Miranda was a piece of good fortune for JPL, for Miranda turned out to be one of the most intriguing objects in the whole Solar System. It is hardly an exaggeration to say that the difficult and ingenious Uranus sector of the *Voyager 2* epic would have been worthwhile for the revelations about Miranda alone.

Miranda is the smallest and innermost of the five known satellites of Uranus, with a diameter of about 500 km. The imaging team at JPL were totally unprepared for the pictures that came back as a series of thousands of digital bits across the enormous gulf between Uranus and Earth. In the words of Larry Soderblom, the assistant leader of the team and a veteran of the *Mariner* missions to Mars: 'It was a bizarre hybrid of every kind of exotic terrain in the Solar System.' Like the other four moons, Miranda turned out to be composed largely of ice mixed with silicate rocks and methane, frozen because of the low temperature prevailing this far from the Sun. But it was the form in which these ingredients were mixed and their strange surface forms which gave the team such food for thought. On the computer-enhanced images they could see a patchwork of sinuous valleys like those found on Mars. There were also areas of grooved terrain, similar to that found by *Voyager* on the surface of one of Jupiter's satellites, Ganymede. Elsewhere, the surface of Miranda resembled the cratered highlands of our own Moon, and there were also giant scarps higher than the Grand Canyon. In the centre of the satellite was a large rocky area shaped rather like a chevron, and two multi-ringed features rather like archery targets bracketed it. Tentatively, the strange appearance of Miranda has been explained in this way: At some remote time in the past, Miranda, then an ice ball with a rocky core, was broken into fragments by a collision, perhaps with another Uranian moon. Then gravitational forces re-assembled the fragments into spherical form, but in a random way, so that some of the pieces settled with the ice side out, and some with the rock showing.

Three of the other satellites show signs of 'geological' activity—a word which should really be restricted to use about the Earth. Oberon and Titania display heavily cratered areas, while Titania also has subsidence features as well as trenches and scarps on a large scale. Ariel, with a diameter of 1,200 km, is most active: it is criss-crossed with valleys and fault scarps. Some of the valleys are flooded with a dark substance, just as craters on the Moon are flooded with lava. In Ariel's case, however, the 'lava' is probably frozen methane which has been darkened over the eons since the formation of Uranus by damage caused by the radiation belts that surround the planet.

Paradoxically, Umbriel is the satellite that gave the watching scientists the greatest reason for scratching their heads. The other satellites show all the signs of considerable internal energy, probably through tidal interactions with each other in the distant past. But Umbriel, while heavily cratered, is dark and inert. It is covered uniformly by craters of between 100 and 200 km in diameter and there is little variation in colour or brightness, save for a white ring 150 km across, which is probably a frost-lined crater. Why Umbriel should be such an inert body is a mystery at the time of writing.

As the *Voyager* sped through its six-hour close encounter with Uranus its instruments detected a magnetic field, comparable in strength with those around the Earth and Saturn. This was not a surprise because the gas giants all appear to have a magnetic field. Earth's field is generated by convection currents in the molten iron core, and similar currents in electrically conducting fluids, presumably ionized gas, are responsible for those possessed

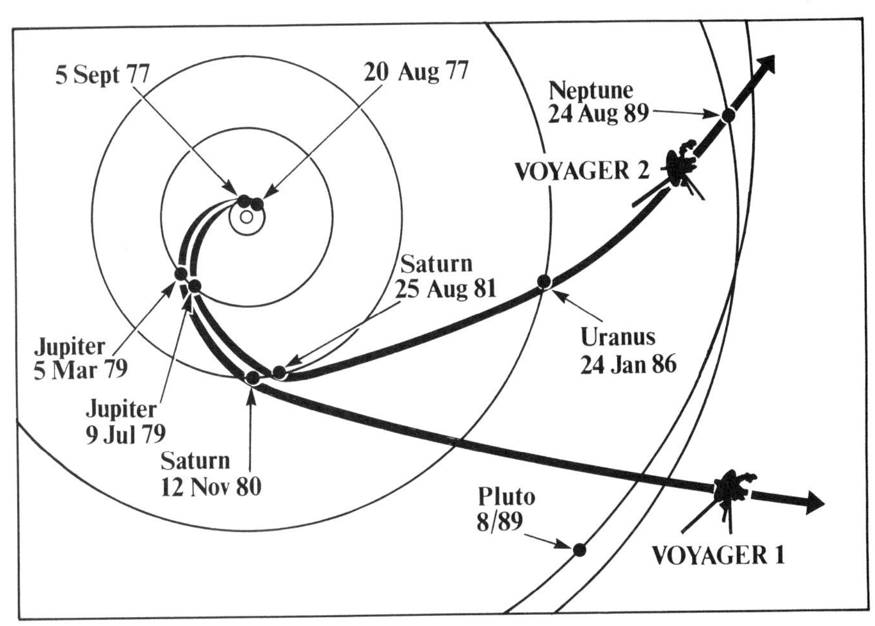

The trajectories of the two Voyagers *during their epic journeys.*

by Jupiter and Saturn. The implication of the discovery is that Uranus too has a body of such a fluid beneath its atmosphere.

But the planetologists believe that a planet forming at the distance of Uranus from the Sun would have been made from silicate rocks with a matching amount of water ice. The model then is of a three layer planet, a rocky core surrounded first by a mantle of water thousands of kilometres deep and topped by several thousand more kilometres of hydrogen and helium in the atmoshere. The 'ocean' would be electrically conducting because the enormous pressure of the atmosphere would break the water down to oxygen, hydrogen and other ions. If the existence of the magnetic field was not a surprise, its orientation was. It turned out to be offset from the axis of rotation by a remarkable 60°, compared with the 11° which causes the Earth's magnetic North Pole to be hundreds of kilometres from the geographical pole.

The offset was fortuitous, because it made it possible to calculate the planet's rotational period, otherwise difficult in the absence of any conspicuous surface features that could be tracked. It turned out to be 17.3 hours. The calculation was possible because, like all planets with a magnetic field, Uranus leaves behind it a 'magnetotail' which could be measured and plotted. Unlike other planets, however, Uranus leaves behind a magnetotail in the form of a corkscrew because of the orientation of the magnetic field, and the turns in space match the rotational period of the planet.

After all these discoveries had been made, there was still new information about Uranus to be distilled from the amazingly successful *Voyager 2* mission, in particular about the rings around the planet, which were discovered as recently as 1977 by Earth-based astronomers. By taking its cameras so close, *Voyager* was able to send back images of the nine known rings 10,000 times better than those taken through terrestrial telescopes. It also discovered two new rings and sent back enormous quantities of data which will enable the mission scientists to sort out the fine structure of the rings. *Voyager 2* appears to have confirmed the theory which had been proposed to explain the origin of the rings, the 'shepherds and sheep' theory.

It is believed that the rings accrete from the debris surrounding the planet, anything from microscopic dust particles to boulders as big as a house, and the pictures recorded two moonlets, inside and outside the largest ring, which is known as the epsilon ring. These tiny bodies are only 30 and 40 km in diameter respectively, and the 'shepherds' of the other rings are probably too small to be resolved by *Voyager*'s cameras, although there is corroborative evidence for the theory from Saturn, where *Voyager 1* found tiny moonlets 'shepherding' that planet's F ring. The rings are as black as coal and it was a considerable technical feat to get any images of them at all. The rocks that make up the rings are probably coated with the same type of radiation-damaged methane that makes up the 'lava' on Ariel. Eight other new minor satellites were discovered and enhancement of the images over the next few years may reveal further rings. For instance, the epsilon ring, which is the widest at 36 km, but which appears to be only 25 to 30 *metres* thick, may actually be composed of at least three rings.

Voyager 2's encounter with Uranus was a complete triumph—a tribute both to the teams of scientists who have nursed the craft more than halfway across the Solar System, and to the craft itself more than eight years after its launch. Its next target is the planet Neptune, more than 4,500 million km from the Sun. After that encounter the little scrap of man-made ingenuity will soar away at a steady speed into interstellar space, still sending back data on the medium through which it is travelling for several years before the signal fades away in the noise of empty space. Any intelligent race with the ability to recover *Voyager* if it brushes by some other Solar System in the remote future will probably marvel at the audacity of a people ready to send out such a messenger to the stars. It may be one of a myriad of such craft wandering through our galaxy, but it is unique to man.

Chapter 11
Chasing a comet

In 1986 a flotilla of spacecraft from Europe, the Soviet Union and Japan took part in a mission which would have been unthinkable when planetary exploration began in the late 1950s. It was not strictly a planetary flight but it borrowed much of the technology which had been developed over three decades and it should be mentioned briefly in the context of exploring the planets.

The target was the much-heralded Halley's Comet. In its highly elliptical orbit this most famous of all comets comes zooming back into the vicinity of the Sun every 76 years before returning to the depths of space and the anonymity of darkness far away from curious humans. It is the only bright, short-period comet that is interesting enough to warrant a visit by human artefacts and that was why three space agencies—but not NASA—decided that the 1986 visitation should be the first in which close-up scientific study should be attempted. Strictly speaking, it was not the first comet to be studied by a spacecraft, because with the aid of a gravity assist from the Moon NASA was able to re-direct an international solar explorer known as ISEE which was in heliocentric orbit to fly through the tail of a comet known as Giacobini-Zinner in 1985. But the International Comet Explorer, or ICE, as ISEE was then re-named, did not make a major contribution to cometary studies. By contrast the international scientific community learnt more from the 1986 visitation of Halley's Comet than they had learned from the previous three centuries of study.

Undoubtedly the most exciting of the little fleet that intercepted Halley's Comet was *Giotto*, the European Space Agency's contribution, which was named after the Italian painter who was the first to produce a reasonable image of the famous visitor. Unlike the other craft targeted on Halley's, *Giotto* was aimed at the 'coma' of the comet and in the event passed 605 km from the nucleus. At that range and at the closing speed of some 68 kms this was a risky course to take, but despite some hundreds of impacts by dust particles being ejected from the 'dirty snowball' that the comet nucleus resembles *Giotto* survived and is still in contact with Earth. *Giotto* was the first interplanetary spacecraft to be launched by the Ariane rocket—largely built in France—which is ESA's competitor to the US Space Shuttle. It was

Above Giotto *is launched as the payload of an Ariane rocket from the European Space Agency base in French Guiana.*

Above right *This picture from* Giotto *shows the nucleus of Halley's Comet from a distance of 25,650 km. Contours show different levels of brightness.*

Right *The* Giotto *spacecraft completes demagnetization tests in Munich after its construction in Bristol.*

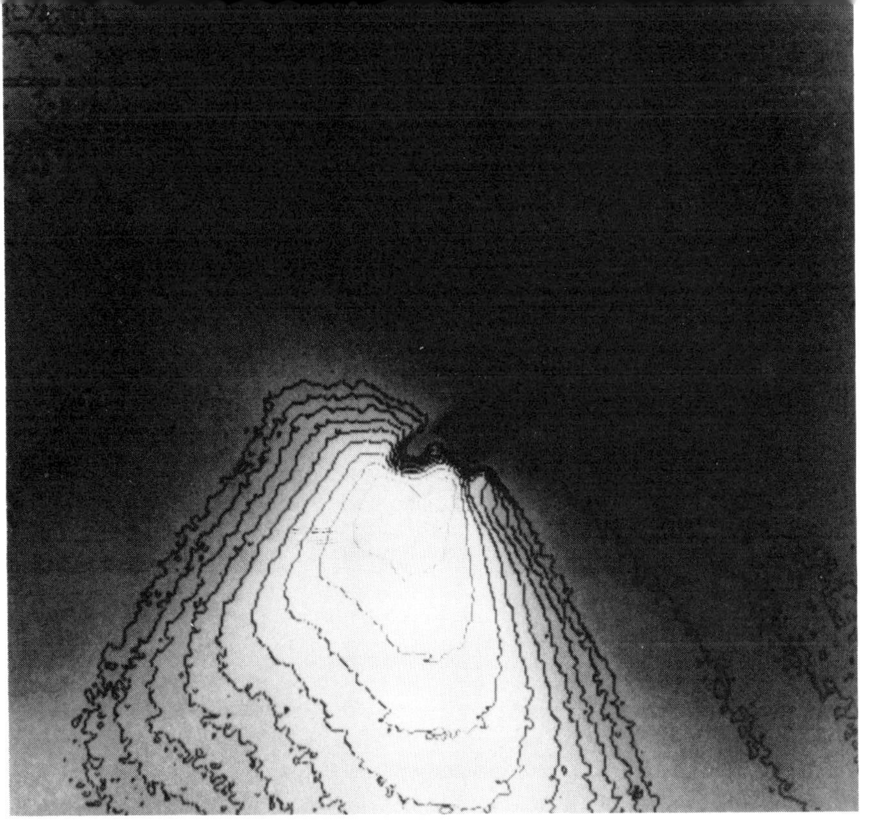

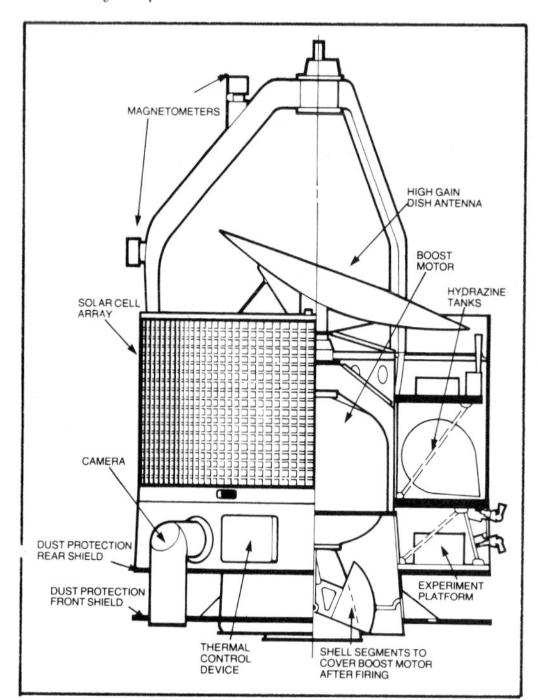

The brilliantly successful Giotto *spacecraft, a product of combined European industry.*

launched from the Kourou base in French Guiana on 2 July 1985, and after three revolutions round the Earth in a parking orbit its own rocket engine propelled it into a heliocentric orbit designed to intercept the comet some eight months later.

It was as technically sophisticated as one might expect from the best of Europe's industry, drawing on three decades of planetary exploration experience. At the heart of the 960 kg spacecraft is what is called a multi-colour camera, designed to send back pictures representing different levels of brightness in the comet's structure by different colours. It is a type of camera which will be common in spacecraft in the future for it is based on the concept of the CCD—the Charge Coupled Device—a system of micro-chips which translates light falling on it into electronic impulses and which is more sensitive and accurate than the advanced TV cameras used up to the present.

In addition to the camera *Giotto* includes spectrometers to analyse the gases in the coma of Halley's, magnetometers, an optical probe for measuring the positions and masses of dust particles and devices for studying the intricate plasmas which surround and influence the comet. To summarize the mission, *Giotto* was a complete success and on 14 March 1986, achieved its rendezvous with the famous comet. The 2,000 pictures that were sent back were not immediately recognizable as a solid body to the man in the street but they had immense significance to the scientists who had planned the operation. They showed that the nucleus of Halley's Comet is a potato-shaped object about 15 km by 10 km, perhaps twice as large as was thought before *Giotto* made its close approach. Issuing from the comet's surface, which was as black as any object in the Solar System, were dust jets which are caused by the intense pressure of solar radiation and which seem bright because they reflect sunlight. The comet was giving off about 60 tons of gas every second and the spectrometers showed that much of it was in the form of ions resulting from the breakdown of water molecules. There were also ions of other elements such as carbon, nitrogen, iron, cobalt, copper and nickel. These findings are very important, for it is believed that comets show an inventory of the original nebula from which the Solar System was formed. By examining the comet so closely, *Giotto* was really looking back into the history of our system and making observations that are impossible anywhere else.

Giotto survived the encounter with the comet, although the scientists at the control centre at Darmstadt in West Germany feared the worst two seconds before the closest approach, when transmission was interrupted. What had happened, it was deduced later, was that dust impacts had caused the high gain antenna to lose its fix on the Earth. Control was regained soon after the comet had been passed and it appears that the damage was slight, thanks to a double dust shield which lessened the effects of impacts which could otherwise have been fatal. *Giotto*'s thrusters were fired soon after the encounter to alter the trajectory so that it will come close to Earth in its heliocentric orbit in July 1990. It may be possible to encounter another comet in the future as the power supply, relying on advanced solar cells, is still adequate.

One of the satisfying aspects of the 1986 campaign of study of Halley's Comet was the extremely successful international co-operation between the various nations mounting expeditions, and also NASA, which supplied the expertise of the Deep Space Network to make accurate targeting possible. The Soviet Union launched two *Vega* craft on 15 and 21 December which, as mentioned in Chapter 4, studied Venus with landers and atmospheric balloons before proceeding to fairly distant encounters with the comet (the name *'VeGa'* comes from the Russian *'Venera'* and *'Gallei'*—RTHalley). Ultra-violet cameras on the *Vega* craft sent back images of the coma of the comet from long range and from the tracking by the Deep Space Network the Europeans were able to establish the comet's orbit more accurately and so ensure the closeness of their own rendezvous. The Japanese, too, co-operated with findings from their two craft, *Suisei* ('Comet') and *Sakigake* ('Pioneer'), which flew past the comet just before *Giotto* and returned plasma and magnetic data and images from an ultra-violet camera.

By chance, the veteran *Pioneer Venus* orbiter, which continues to send back data about the planet, was in a position to use its photopolarimeter to obtain interesting ultra-violet images of the comet, which certainly became the

Left *The* Vega *spacecraft, a modified* Venera *which studied both Venus and Halley's Comet.*

Right *The* Vega *instrument package which descended through the Venusian atmosphere.*

Below *The sophisticated scan platform of* Vega, *bearing the instruments which studied the comet.*

most studied object in the Solar System after the Sun and the planets. There were even data to be obtained from some of the *Pioneer* series, *6, 7, 8* and *9* which had been in heliocentric orbit at roughly the same distance as the Earth since the late sixties. Altogether, the Halley's Comet visitation of 1986 was a shining example of the way in which scientists of all nations can co-operate fruitfully in the peaceful exploration of the Universe.

Chapter 12
The future

Although the hectic pace of the sixties and seventies has slowed down, there is still a great deal of activity in planetary exploration. To begin with, *Voyager 2* is still *en route* to the planet Neptune, which it will fly past in August 1989. That is a certainty—the laws of celestial mechanics have the little spacecraft in their grip and the mission planners have only a slight amount of control over the direction of travel. At the time of writing it had not been decided whether to go for the 'polar crown' mission, flying only 1,300 km above Neptune's north pole, with a close fly-by of the largest moon, Triton, or for a more distant pass aimed at the discovery of any rings that the planet may possess. The former was apparently the more likely of the two alternatives to be chosen.

It is to be hoped that *Voyager*'s cameras and instruments, radio and computers are in excellent working order in 1989 and that this sector of the mission is as successful as those that have ended. After Neptune, *Voyager 2* will continue flying in a straight line out of our Solar System in the direction of Sirius, the Dog Star. Although its speed relative to the Sun is more than 56,000 kmh, it will not encounter another stellar system until some millions of years have passed. To make sure that the craft's origin is not a mystery if some other intelligent race happens on it in space at some time in the distant future, *Voyager 2* has been fitted with a video disc giving a whole selection of sounds and views from Earth, including music and great literature of various languages.

Voyager 1, which is still transmitting data on the interplanetary medium, ended its planetary mission with the encounter with Saturn. But it too has been given an impulse which will take it out of the Solar System. In this case the trajectory is an exciting one scientifically because it is not staying in the plane of the ecliptic—the plane in which all the planets revolve around the Sun—but is going up and out of it. The region above the plane of the ecliptic has not been explored thoroughly before and there should be some interesting findings before *Voyager 1*'s signals become too weak to be received as it soars towards the constellation of Ophiucus.

The *Challenger* Space Shuttle disaster in January 1986, has brought toan abrupt halt the continuation of the US planetary programme. The next mis-

Galileo, *the next spacecraft to be sent to Jupiter, is now unlikely to be launched from a Space Shuttle, as depicted in this artist's impression. The pointed probe will penetrate the Jovian atmosphere.*

sion, which was to have begun with a Centaur launch from the *Challenger* in May 1986, was Project *Galileo*, an ambitious attempt to explore Jupiter more comprehensively than was possible with the *Voyager* fly-by approach.

Galileo will adopt a technique which is common with communication satellites in Earth orbit but has not previously been used in planetary studies. Part of it spins at three revolutions per minute, providing stabilization, but the non-spinning section will furnish a stable platform for cameras and instruments pointing at the planet and its satellites. The spacecraft will go into orbit round Jupiter, so providing prolonged opportunity for study of the atmosphere and environment of Jupiter, and it will also pass close to the four Galilean moons on ten occasions during its 22-month planned mission. Encounters with the satellites will be as close as a few hundred kilometres.

The final refinement is that 150 days before reaching Jovian orbit the mother craft will release a probe which will plunge in to the atmosphere. The probe will sample the gases of the atmosphere for the first time and transmit data on temperature and pressure for a precious few hours while it descends

Ulysses, *a joint European-US probe to study the poles of the Sun through a gravity assist by Jupiter. It is another Shuttle payload which will probably have to be switched to an expendable booster.*

behind a heat shield and then beneath a parachute. *Galileo*, which has been planned with the co-operation of West Germany, is the finest product of the talented team at the Jet Propulsion Laboratory and, with advanced computers and instruments added to their many years of experience, *Galileo* promises to be a fruitful mission.

A comparable enterprise by the European Space Agency has also been delayed by the destruction of the *Challenger*. This is *Ulysses*, an ingenious mission designed to study the poles of the Sun, which are not accessible by ordinary propulsion methods in the plane of the ecliptic. *Ulysses*, another payload for the new version of the Centaur, will be launched *outwards* from the Sun, loop round Jupiter and back towards the Sun with the extra impetus from the planet's gravitational field enabling it to pass over the Sun's north pole. There were to have been two craft in this unique 'International Solar Polar' expedition, but NASA cried off for budgetary reasons and Europe was left to go it alone.

The *Challenger* disaster has made a big change necessary in the method of launching these two missions, as well as in delays to their timetable. Critics of NASA have won the argument that the Centaur, fuelled as it is with liquid hydrogen, is too dangerous a payload for the shuttle cargo bay. As a result, both *Galileo* and *Ulysses* were removed from the cargo manifest for the shuttle and may now have to be launched by an unmanned Titan rocket with the Centaur as the upper stage. An alternative plan presumes a swift resumption of shuttle flights and the use of the new IUS rocket to launch *Galileo* from the shuttle cargo bay in November 1989. The version of the Titan which will become available is actually being built for the US Air Force, so at the time of writing it was not known when they might be available to NASA and or when *Galileo* and *Ulysses* will be launched.

America will be going back to Venus, perhaps as early as 1988, with a spacecraft which will map 90 per cent of the planet's surface by radar with 1 km resolution. The Venus Radar Mapper, to be named *Magellan*, will produce the definitive map of the planet, not much less than thirty years after man began to explore Venus by remote control.

The first of a series of new 'Planetary Observers' planned by NASA will be the *Mars Observer* in 1990, to be placed in a low polar orbit round the Red Planet. It will conduct a detailed study of the constituents of the surface and of the atmosphere.

Mariner Mark 2 missions are just a gleam in the eye of the JPL team at the moment but if they are approved they may include a joint mission with ESA to place an orbiter around Saturn with a probe being dropped into the atmosphere of the satellite Titan, and a fly-by of an asteroid on the way to a rendezvous with a short period comet. An unmanned rover vehicle roaming across the lava plains of Mars is also a remote possibility. All these missions will be hit by the *Challenger* disaster, since at least a year and possibly rather more has been taken out of the timetables for shuttle missions. More important, they would all be Centaur payloads, so they will probably all be removed from the shuttle programme unless the Centaur itself can be made safer, or the new IUS rocket proves successful.

It is in this context that last year's report by the National Commission on Space must be viewed. The Commission's report recommended many desirable aims for the US national space programme but its financial implications were not followed up sufficiently. Time will tell whether lunar bases and manned expeditions to Mars will be sufficiently attractive to persuade future administrations and taxpayers to provide the resources necessary.

Similar concerns must be felt in the Soviet Union, where the space budget takes a relatively higher percentage of the gross domestic product. There has been no mission to Mars since 1973 and no dedicated flight to Venus since *Venera 15* and *16* in 1983, although the two *Vega* craft dropped probes into the Venusian atmosphere on the way to their Halley's Comet encounter. Further *Venera* missions may be in the planning stage and it is known that the Soviet Union will be returning to Mars in 1988 after a lapse of fifteen years. The mission will employ two of the large Proton boosters to place twin spacecraft into orbit round Mars and land small instrument packages on the two moons, Phobos and Deimos. This mission will be an order of magnitude more complex than anything the Soviet Union has attempted in the past. It will be a co-operative effort with several different countries, including France and West Germany and will challenge the supremacy of the United States in high technology planetary exploration. If it succeeds and is followed by similarly complex Soviet missions to the Moon and to other planets it will open a new era which will probably end with the first manned expedition to Mars in the early 21st century.

This would be a further order of magnitude more difficult than the 1988 lander mission to the satellites of Mars, but the Soviet Union seems capable of making this huge leap. With America in its present mood of suspicion of space exploration, this expedition would almost certainly be a Soviet one, although the space community throughout the world would favour an international collaboration in this most important of all space missions. It might very well be the catalyst which could turn man's more bellicose instincts into a constructive urge to explore the Universe as a species and not as a series of separate nations.

Even in the context of national rivalries, the years since 1957 have seen some remarkable achievements in the field of planetary exploration. We now know far more about the planets than our ancestors could ever have dreamed was possible. It is to be hoped that future strides will be just as remarkable and rapid and will help to divert man's creative energy and resources to peaceful and gainful pursuits.

Glossary

Alpha back-scattering A method of analysing rocks by subjecting them to alpha radiation and detecting the responses of different elements.
Apoapsis The high point of a spacecraft's orbit round a planet.
Apogee The high point of a spacecraft's orbit round the Earth.
Aposelene The high point of a spacecraft's orbit round the Moon.
Beta particles Nuclei of the element helium.
Doppler tracking Establishment of a spacecraft's position and velocity by measuring the change in frequency of radio signals received from it as it travels through the Solar System
Escape stage The last stage of a rocket which imparts to the payload the velocity necessary to escape from Earth's gravitational field.
Escape velocity The speed which a spacecraft or a particle needs to attain in order to escape from the gravitational field of a planet or star. In the case of the Earth the velocity is 11.2 km per second.
Figure The precise shape of a planet and how it differs from a perfect sphere.
Greenhouse effect The heating of a planet's atmosphere caused by a gas such as carbon dioxide preventing the energy of reflected sunlight from escaping into space.
Heliocentric orbit An orbit in the Solar System with the Sun at its centre.
Infra-red Radiation of longer wavelength than visible light.
Ionosphere The region round a planet in which the gases of the atmosphere are ionised.
Launch window The period during which a spacecraft must be launched in order to reach a particular objective, dictated by the relative positions of the Earth and the target planet.
Magnetometer An instrument which detects magnetic fields in space.
Magnetopause The boundary between a planet's magnetic field and the solar wind.
Magnetosphere The region round a planet in which its magnetic field predominates over the solar wind.
Mascon A concentration of heavier material beneath the surface of a planet or the Moon, causing anomalous gravitational effects.
Periapsis The low point of a spacecraft's orbit round a planet.

Perigee The low point of a spacecraft's orbit round the Earth.
Periselene The low point of a spacecraft's orbit round the Moon.
Photopolarimeter An instrument that measures the intensity and polarity of light, enabling an image of a planet to be built up without a camera.
Plasma A state of matter in which highly energetic atoms have been stripped of their electrons and so are electrically charged.
Retrograde orbit An orbit in which a satellite is travelling in a direction opposite to the rotation of its parent planet.
Retro-rocket A rocket used to slow down a spacecraft, in order to place it on the surface of a planet, or in orbit round the planet.
Solar wind The outpouring of charged particles from the Sun which is encountered by all the planets.
Spectrometer An instrument for identifying elements by detecting their characteristic emissions in the electro-magnetic spectrum.
Ultra-violet Radiation of shorter wavelength than visible light, which is emitted by many heavenly bodies.
Vernier engine A small rocket motor used to steer a booster during powered flight or a spacecraft during landing on a planet.

Index